MATH MODELING

COMPUTING & COMMUNICATING

Karen M. Bliss
Department of Applied Mathematics,
Virginia Military Institute, Lexington, VA

Benjamin J. Galluzzo
Department of Mathematics,
Shippensburg University, Shippensburg, PA

Kathleen R. Kavanagh
Department of Math & Computer Science,
Clarkson University, Potsdam, NY

Rachel Levy
Department of Mathematics,
Harvey Mudd College, Claremont, CA

DESIGN
PlusUs
www.plusus.org

PRODUCTION
First Edition 2018

siam. | Society for Industrial and Applied Mathematics

CONTENTS

INTRODUCTION

This guide is intended for students, teachers, prospective teachers, teacher educators, and anyone who wants to use software, technology, or computer programming during the math modeling process. While it is mainly geared toward high school and college students and teachers, some of the information may be useful for teachers in lower grades. We have not written a text for a modeling course; this guide provides a way for modelers to jump in and get their feet wet.

A mathematical model is a representation of a system or scenario used to gain qualitative and/or quantitative understanding of some real-world problem, to predict future behavior, and to guide decision-making. Math modeling now appears in K–12 and college curricula because modeling is recognized as a critical tool used to investigate challenges facing our local and global communities. We want to open the door for people just getting started and looking for ideas about why, when, and how to use tech tools. We will briefly review the process of math modeling in the next section, but if you need more information about it, you may want to check out the guidebook "Math Modeling: Getting Started and Getting Solutions" [4]. We'll refer to it throughout this handbook as *M2GS2*.

Often when we model real-world problems, we can't find a solution without using some form of technology. Luckily, we have many computational tools at our fingertips which can be used, along with increasingly accessible large data sets that describe human behaviors and the world around us. In this guide, we focus on ways spreadsheets, mathematical software, and programming languages can be used in the modeling process. Within each of those categories are both open-source and proprietary software packages for you to explore. The open-source options are usually low cost or free, while the proprietary ones often have free or reduced price options for students.

The more you use a software tool, the more skilled with it you will get. Many mathematical software tools can perform like a graphing calculator, but also have their own programming language that allows you to write functions or use built-in computational tools. At first, you may be frustrated and realize you are not using the right commands or don't know how to make something work, but within that productive struggle you are *learning* how to use those tools. It will be easier the next time, so stay the course!

What matters most as you get started is that, after you develop the big picture idea for your model and some of the details, you take some time to choose the appropriate software for the task at hand. This book will help you make those choices. If you find yourself trying too hard to make a tool work for your task, maybe something else would be a better fit. It might be worth taking some time to explore what is available.

Each of the next four chapters will focus on one of the following ways that software is used in modeling: statistical analysis, data visualization, computations, and simulations/ programming. We also include a chapter with helpful information about the popular software tools that we use throughout the book.

It should be noted, however, that you may end up navigating the book in a nonlinear way; be prepared to switch from one chapter to another as needed. Here's one example of how you might navigate the modeling process with technology (and with this book):

> You may start answering a modeling question, and determine that you need to get some data. You do an internet search and locate data you'd like to investigate. First you decide to plot the data (Visualization chapter). You find that your initial choice of visualization is not helpful, and that you need to sort your data and decide what to do about missing data (Statistics chapter). Then you make a better plot (Visualization chapter) and realize that there are several outliers. You decide what to do with the outliers (Statistics chapter), and then you decide you want to fit the data with a curve (Statistics chapter). You then use this curve in the model to extrapolate values forward in time (Computations chapter). After some more work, you polish up your figures for your final presentation (Visualization chapter).

As you go through this guide, it may help to be online and also to have a statistics resource nearby. We will use lots of examples to demonstrate principles, highlight best practices, and outline some *Do's and Don'ts*.

1.1 OVERVIEW OF THE MODELING PROCESS

To start, we give an overview of the components of math modeling, including a brief review of some terminology.

A mathematical model takes *inputs* (often referred to as *independent variables*), manipulates those, and returns an *output* or outputs (i.e., *dependent variables*) that help us make decisions. You might make some of the quantities constant when you evaluate the model and then later go back and change those quantities to see how the model output changes. Those quantities are called model *parameters*.

As you review the components of the modeling process, it's important to remember that this isn't necessarily a sequential list of steps; math modeling is an iterative process, and the key steps may be revisited multiple times, as we show in Figure 1.

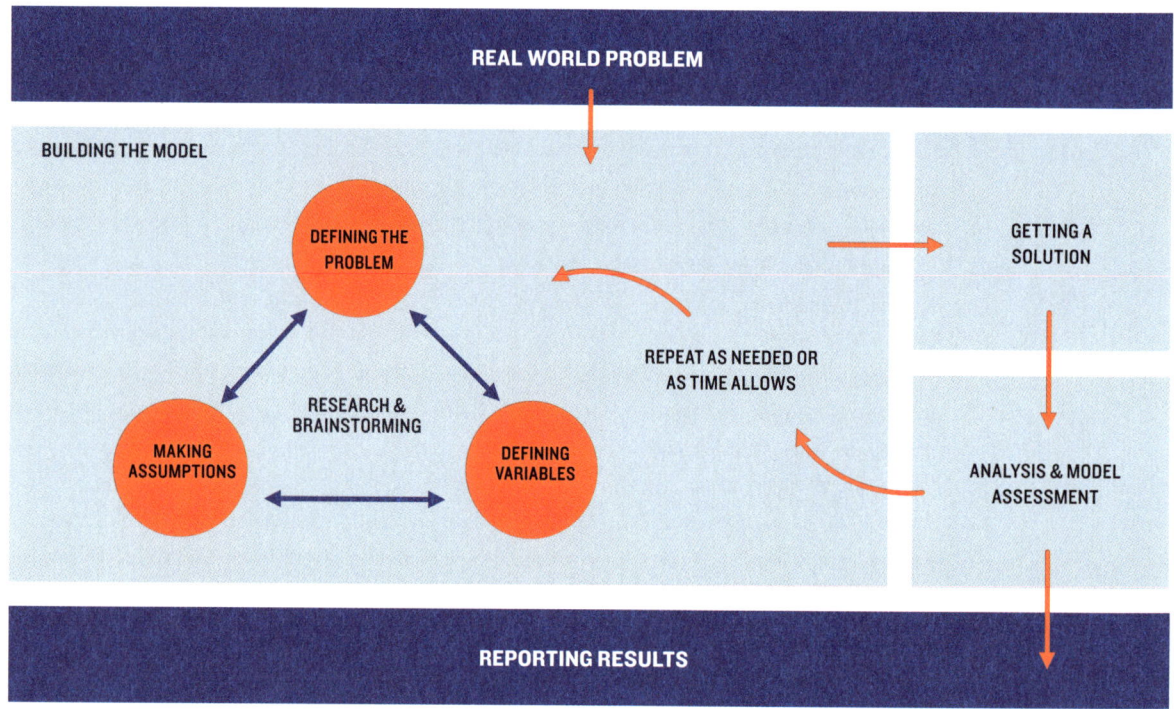

Figure 1: The math modeling process [4].

- **Defining the problem statement.** Real-world problems can be broad and complex. It's important to refine the conceptual idea into a concise problem statement which will indicate exactly what the output of your model will be.

- **Making assumptions.** Early in your work, it may seem that a problem is too complex for you to make any progress. That is why it is necessary to make assumptions to help simplify the problem and sharpen the focus. During this process you reduce the number of factors impacting your model, thereby deciding which factors are most important.

- **Defining variables.** What are the primary factors influencing the phenomenon you are trying to understand? Can you list those factors as quantifiable variables with specified units? You may need to distinguish between independent variables, dependent variables, and model parameters. In understanding these ideas better, you will be able to both define model inputs and create mathematical relationships, which ultimately establish the model itself.

- **Getting a solution.** What can you learn from your model? Does it answer the question you originally asked? Determining a solution may involve pencil and paper calculations, evaluating a function, running simulations, or solving an equation, depending on the type of model you developed. It might be helpful to use software or some other computational technology.

- **Analysis and model assessment.** In the end, you must step back and analyze the results to assess the quality of the model. What are the strengths and weaknesses of the model? Are there certain situations in which it doesn't work? How sensitive is the model if you alter the assumptions or change model parameter values? Is it possible to make (or at least point out) possible improvements?

- **Reporting the results.** Your model might be awesome, but no one will ever know unless you are able to explain how to use or implement it. You may be asked to reduce bias in your results or to be an advocate for a particular stakeholder, so pay attention to your point of view. Include your results in a summary/abstract at the beginning of your report.

As you report your results, it is important to remember that your model does not dictate reality. As you make use of technology, be sure not to perpetuate myths such as "the computer said it, so it must be true." Instead, try to convey that you are providing your best understanding, given the affordances and limitations of the model you created.

1.2 WHILE MODELING, WHEN MIGHT YOU USE TECHNOLOGY?

When building your model, you will often get to the point where you need to do some number-crunching. In the following example we provide a collection of scenarios you may encounter when working through the modeling process, and provide a technologically driven approach that will push your project forward.

Lunch Crunch example

Suppose you have been assigned a task to find affordable and nutritious—yet also delicious—lunch options for a school district.[1]

- *Getting started:* Let's imagine you are brainstorming modeling approaches for the Lunch Crunch question, focusing first on making sure the lunch provides the right number of calories. You decide to make sure you're providing enough calories (so students are nourished and ready to learn), but not too many (so the school lunch doesn't contribute to the obesity epidemic). Usually brainstorming involves asking questions, such as: How many calories do students need in a day? How many calories are in foods they like to eat? How much do healthy, tasty foods cost?

> In this early stage of model development you might just pick up a calculator and try a couple of numbers to make sure the output of your proposed model is at all sensible before you take the time and effort to implement the model with computing software and/or simulation software.

• *Building a model:* In trying to find out how many calories students need, let's say you run across a pre-existing model scientists have built to help them answer that very question, and you decide to use it as the base for your own model. The revised Harris–Benedict equation [5] assumes that you can get a reasonable estimate for the total number of calories needed per day if you have three pieces of information about an individual: (1) activity level factor (ALF), (2) basal (resting) metabolic rate (BMR), and (3) thermic effect of food (TEF). The basic model equation is

$$\texttt{CALORIES = ALF*BMR + TEF.}$$

The value of TEF is often approximated by

$$\texttt{TEF = 0.1*ALF*BMR,}$$

so the model equation is

$$\texttt{CALORIES = ALF*BMR + 0.1*ALF*BMR}$$

or

$$\texttt{CALORIES = 1.1*ALF*BMR.}$$

Now you want to investigate how changes in ALF and BMR can impact the caloric needs of individuals.

In the Harris–Benedict equation the activity level factor ALF takes into account how a person's caloric needs change with respect to their activity level. This is often a factor between 1.2 (for people who are not active) to 1.9 (for very active people). Since there is a range, you will probably want to be able to examine how the output changes when you use different ALF factors.

There are two associated equations for computing BMR [5]—one for men and one for women. You might decide to average the model parameters[2] to get a simpler single model for BMR of both, as follows:

```
BMR = (88.362 + 447.593)/2
    + ((13.397 + 9.247)/2 * (weight in kg))
    + ((4.799 + 3.098)/2 * (height in cm))
    - ((5.677 + 4.330)/2 * (age in years)).
```

■ MEN (LEFT)
■ WOMEN (RIGHT)

As you can see, this model has several variables that are different for each student. Engaging with software is appropriate here because it allows you to easily swap out values for ALF, weight, height, and age and then immediately see the change in output. The idea is to vary the values in a systematic way so that your explorations are easier to automate and save than if you used a calculator. We'll revisit this idea in the Computations chapter.

• *Simulating data:* You now have a model for caloric needs based on characteristics of different people. Now you can test out the model on many different people, such as all the students in a high school.

> One way to do this is using technology to simulate a whole high school of imaginary people, where you generate the data you need for each person based on assumptions about that population. We'll show you a few ways to do this in the Programming and Simulation chapter. Later you might also simulate different lunches to see how many of the students get their caloric needs met.

• *Analyzing and summarizing data with statistics:* You have generated a student population for use in your model. You want to double check that this simulated data has characteristics similar to a real high school population. For example, what is the average caloric intake of your simulated population? Does it roughly agree with published numbers?

> You can use ideas from the Visualization and Statistics chapters to check and justify your modeling decisions.

• *Sharing your work:* Once you have developed a model and found some results, you're ready to find an effective way to share your work with others.

> Graphs, charts, and tables can help others understand your work, and you can find ideas in the Visualization chapter for communicating your solution, including its strengths and weaknesses.

In the Lunch Crunch example, the calculations (a lot of addition, subtraction, and multiplication) provide a strong hint that technology will be useful. Don't be discouraged if this isn't as obvious when you look at other modeling scenarios. It can really help to be flexible and make sure to step back periodically and make sure you're using the right technology for your model.

In this book we have chosen a few tools to provide concrete examples, but of course there are many options. We have highlighted Excel, MATLAB, Python, and R because they allow us to show a variety of approaches and because they are tools that we have seen used by high school students that also have a place in industry. Other tools, such as Geogebra (for things like slope fields) and Netlogo (for simulations) have been suggested as useful software students might want to explore.

As you gain experience with technology and software, you will undoubtedly become more comfortable using some tools more than others. But be careful; if you have a hammer, you

FIGURE 2: TECHNOLOGY MIGHT BE USED THROUGHOUT THE MODELING PROCESS

WHILE BRAINSTORMING...

– You realize there is little or lots of data available. Can technology help you sort through or create information you may need to define the problem?

– Did you find a pre-existing model that could help you get values to use in another part of your model? Can technology help you evaluate the model or implement it from scratch?

WHILE BUILDING THE MODEL...

– You find some data. Can a quick visualization help you get ideas about how to build your model?

– You need data and can't find it. Can you simulate it?

– You need to use or build a function/submodel that will connect to other parts of your larger model. Can you code it up?

– You evaluate a function over a large data set to get values you need. Can technology make this easier to do or repeat several times?

WHILE GETTING A SOLUTION...

– Do you need some statistical computations such as mean, linear regression, or other curve-fitting?

– Is your model actually a simulation of a real-life scenario? Could you write a program and calculate some quantity over and over, for example over a given time period?

– Do you need to solve a complicated equation, system of equations, or optimization model using a software tool?

WHILE ANALYZING YOUR RESULTS...

– Do you need to vary a model parameter over a range of values to understand how sensitive your model is to that value?

– Would a visualization help you know whether you have a useful solution?

WHILE REPORTING YOUR RESULTS...

– Can technology can help you visualize your results?

– Can technology can help you validate your model by comparing it to some pre-existing data or by evaluating your model with known model parameters and outcomes?

might be tempted to treat everything like a nail. All modelers have to decide whether to use a familiar tool (which may not be the best tool for the job) or take the time to learn something new. Be aware that there are often many ways to use technology to get a solution! As we discuss some of the tools available and how to use them, we will try to point out when you might be tempted to use technology inappropriately and how that might get you into trouble.

As you can see, there are numerous ways in which technology can be used to advance the modeling process. Sometimes you are not focused on technology, and suddenly you realize you will need it (or not!). Our hope is that this guide will point you towards choices and methodologies for using software (when appropriate) at any stage of the modeling process, as shown in Figure 2.

It is important to note that software may complicate a model when a simpler approach is sufficient. As you gain experience with both modeling and software you will learn from those sorts of situations. You will also learn valuable programming and technology skills that may be used in school or at work.

1.3 HOW TO USE THIS GUIDE

We aim to provide examples that illustrate when and how to do computations, statistics, visualization, and simulations using software. The examples are motivated by questions that arise during the modeling process, similar to the Lunch Crunch question we discussed earlier. In each chapter, we will explain when and how a need for software may arise and then show how a modeler might think through their technology choices. Flowcharts such as the one in Figure 3 help visualize the decision processes.

The real world can be messy and complicated, which is why it's so important (and fun!) to do mathematical modeling. We emphasize that there may be other questions that you ask yourself and a variety of other technology approaches not even mentioned here that could indeed help you. However, this book will provide direction for how choices can be made in terms of using the right computational, statistical, visualization, or programming tools while creating your mathematical model.

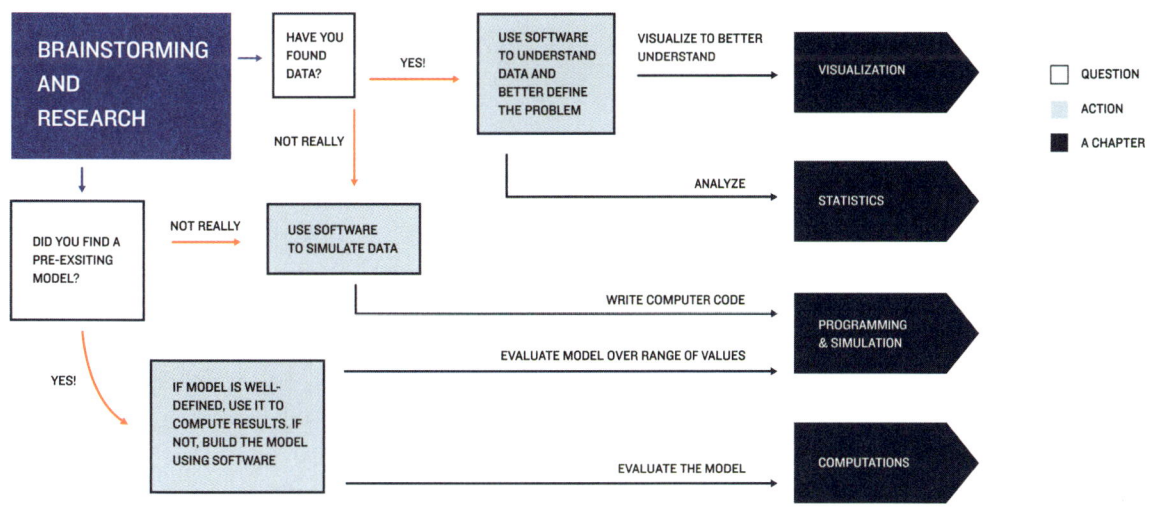

Figure 3: Decision flow chart to illustrate how questions may arise during the brainstorming process and which chapter in the book may help (if you decide that you are not going to do a quick informal calculation).

ENDNOTES

[1] This is a shortened version of the 2014 M3 Challenge problem [1]. The full "Lunch Crunch" problem statement appears in Appendix A.

[2] Averaging model parameter values can be a risky—but sometimes necessary—venture. Make sure you mention this modeling choice when you communicate your model, and revisit the choice if time permits.

STATISTICS

Statistics appear in the news and in advertisements as a way to quantify or summarize information, case studies, and data. Statistics are often used to try to make a convincing argument using data that was collected with a purpose in mind, while other times they are used simply to understand a situation or analyze human behavior. We live in a time where massive amounts of data are being collected from sensors, satellites, and even cell phones. Technology makes even complex calculations with large data sets possible. When it comes to modeling, statistical analysis of data associated with a real-world problem can provide great insight into model building and possible solutions.

Before we get too far into the discussion of how we can get started using statistics in modeling, we need to reiterate that this is not a textbook. That is, in this chapter we will not discuss statistics in the depth of a class or text. Rather, we will focus our discussion on some of the choices modelers make in order to use ideas from statistics. With this in mind, we open the chapter by sharing ideas that everyone, regardless of statistical background, can use. At the end of the chapter, we will highlight a few concepts that require more experience in statistics. We hope that this chapter can serve as a jumping off point for future study of the subject.

When discussing statistics, the term *population* refers to the entire group of things we're interested in studying. A *sample* is a portion of the population. Often, we aren't able to collect information from the entire population. Even when we have access to all the data, sometimes we need to collect data from a continuous process in a discrete way. For example, if I want to track a patient's blood pressure over time, even if I put a blood pressure monitor on the patient, it isn't possible to save the data at every instant—that would be an infinite amount of data! Instead, monitors and sensors are programmed to collect data at discrete (individual) times. When these times are really close together, it can be almost as good as having all the data. In this example, we would identify the population as the patient's blood pressure over all time; the sample is the discrete set of information we were able to collect from the patient when we took the readings.

A *statistic* is a single number used to represent something about a set of data. For example, students often will calculate mean, median, and mode for a given set of numbers.

To calculate statistics, we *sample* from a population to understand its characteristics. For example, in order to get a sense of how many fish there are in a lake, we don't want to catch all the fish. Instead we could put tags on 100 fish. Then we know the fraction of fish tagged is 100/(total population). Later if we collect a sample of 50 fish and find that 2 out of the 50 fish caught have tags, we might assume that 1% of the fish have tags. This means 100/(total population) = 2/50 and we can solve to estimate that the total population of fish is 2500.

If we write this example as an equation, we could write the relationship with variables. Notice that we can have variable names that are words, so hopefully anyone can understand what numbers the variables represent:

$$\frac{\text{(number fish tagged in population)}}{\text{(total fish population)}} = \frac{\text{(number tagged fish caught)}}{\text{(total fish caught)}}.$$

Now to write the equation as a function, we can think about what should be the input and the output of the model. We want to estimate the total fish population, so that can be the output. Let's solve for that variable and rewrite the equation with that number on the left-hand side:

$$\text{total fish population} = \text{(number fish tagged in population)} \times \frac{\text{(total fish caught)}}{\text{(number tagged fish caught)}}.$$

We pull out a sample of fish and count how many fish in this population sample are tagged. This ratio is an input to the model, and the numbers will depend on the situation in the lake we are studying.

There is one more number in the model: we get to decide how many fish to originally tag out of the whole population. In our example, we set the number to be 100. Maybe we decide later that this is too many or too few for a new lake that we want to study, so we change that number to 10 for a small pond or 1000 for a very large lake. Changing the number of fish we tag impacts how well we will estimate the fish population, but otherwise doesn't change the structure of the model. So this number, the number of fish we choose to tag, is called a parameter.

Model parameter: There are multiple interpretations and uses of the word parameter. Statisticians sometimes talk about them as population characteristics. Here we will talk about model parameters as numbers (usually at least temporarily constant) that we are going to choose or estimate.

The model above was created by imagining a way to collect data. When it comes to creating models, you will need to decide whether data is valuable to your model and also how you might use it to build your model. In this chapter we will use some particular data sets to discuss ways you might use technology to assist your statistical analysis. In particular, we'll focus on data used in the 2017 M3 Challenge problem [2] about how the U.S. National Park Service might adapt to sea levels rising as a result of climate-related variability and extreme events. The National Park Service preserves natural and cultural resources (often large amounts of land with interesting natural features) for the public's enjoyment, education, and outdoor recreation. The problem statement appears in Chapter 6, and the problem with data links is available online here:

https://m3challenge.siam.org/archives/2017/problem

I have a modeling question... Should I just launch into data?

You will begin modeling by brainstorming and developing a concise problem statement, and in doing so you will likely be digging around to see what information already exists about the underlying topic of your model. While digging, you may seek out or come across data pertaining to your ideas. Most modeling problems are open to multiple approaches. This means many different data sets could be used to build the model and provide insightful solutions. As the modeler, you must decide which data sets to use and how to use them.

The methods used to collect the data will affect what information you get and the conclusions you draw, but this topic is most appropriate for a statistics course. Here we'll assume you are using data someone else has collected. When you find data to use, instead of justifying your data collection method, you can quickly explain why you believe the data source is trustworthy and why your choices of data are useful, accurate, and representative of the phenomenon you are modeling.

The National Park Service (NPS) problem suggests many data sources to choose from, including mean sea level, heat index, hurricanes, wildfires, temperature, air quality, and visitor statistics. The problem also provides links to websites for the NPS, the Environmental Protection Agency (EPA), the National Oceanic and Atmospheric Administration (NOAA), and related scientific papers.

Should you look at everything? No way. Even if you had enough time, that isn't the best way to use it. First you should brainstorm about what general approach you want to take. Those big ideas will drive the data choices. As you decide which data to use, document your reasoning for your choices. This information will make your solution more understandable for your audience (and trustworthy if you made sensible choices). Once you have a few ideas, you might want to take a quick look at the data to see which idea has the most data support.

Now I'm ready for data. How do I access it?

In order to peek at the data, you'll need to have a program that can open it and display it in a usable way.

One way to get data is to "Cut" and "Paste" it into a spreadsheet. This can be a quick way to copy data that you have found. Unfortunately, sometimes when you paste, the formatting isn't useful for one reason or another (everything gets lumped into one cell, for example). In Excel, you might consider using the `text to columns` command to reformat the data in a useful way.

Make sure you carefully look at your data at this point; missing data can lead to data appearing in the wrong column.

The cut-and-paste method might not work if data is already in a file. Data can exist in a variety of file types (e.g.,.txt, .xlsx, .csv), which tell you how the data is stored (e.g., .csv at the end of the filename indicates comma separated values).

For example, when downloading the visitor data from Acadia National Park, you can see in Figure 4 that you get a pop-up window asking if you want to save the data in different file formats.

Most software can read and export all of these formats, so it's safe to choose any of them (but you can do a quick internet search to see which file types are preferred by your software of choice). Software packages often have their own format, so after you import the data you may end up saving it as an entirely different kind of file, with something after the dot in the filename to let you know which kind.

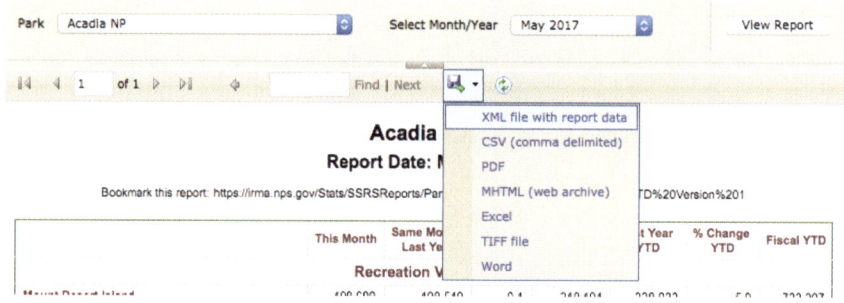

Figure 4: Downloading National Park Visitor Data. Data can be found at [2].

I got my hands on data! Now what?

There are many possible directions you might take to understand your data and the information it can provide. We've organized this chapter by grouping possible actions into three main categories, as shown in Figure 5.

Notice that you might move back and forth between these actions. You should always start by classifying your data. After that, you might sort your data and subsequently visualize it. Once you see it, you might notice some interesting features that prompt you to analyze it with the appropriate statistical tools. Alternatively, you might notice that the data would reveal more information if organized differently, in which case you might reorganize and then visualize it again.

In the following sections, we'll discuss each of the action categories (organize/ prepare, visualize, analyze/examine), but you should be aware that you might not always use them in this order. You'll also see lots of examples along the way to help make sense of the specific actions.

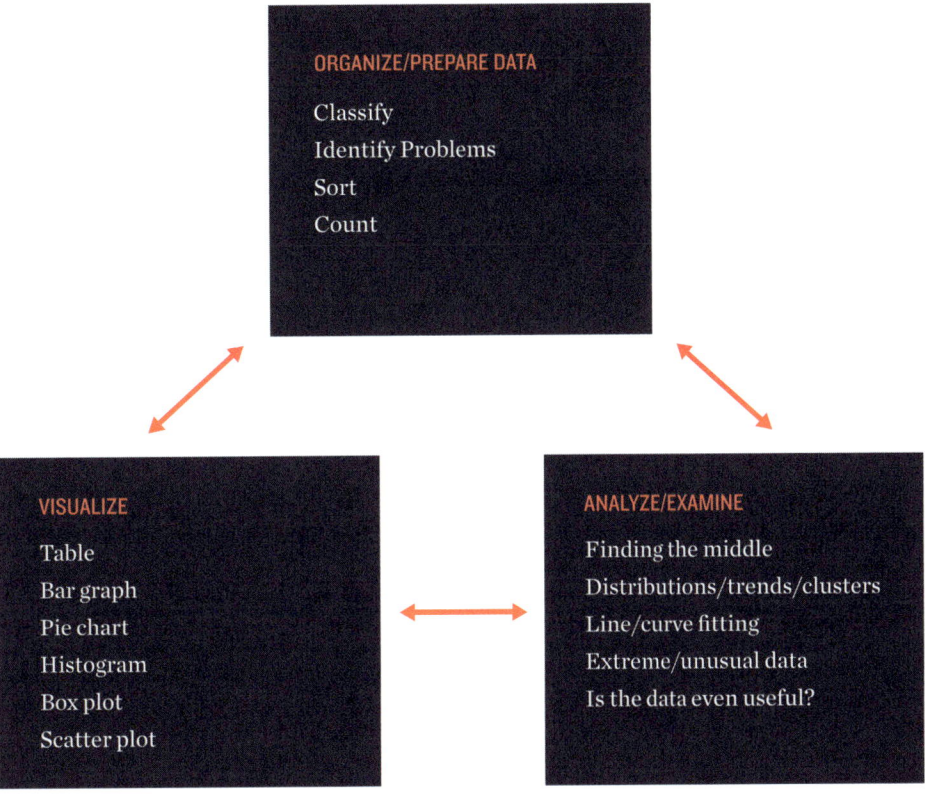

Figure 5: Action categories for starting to understand your data.

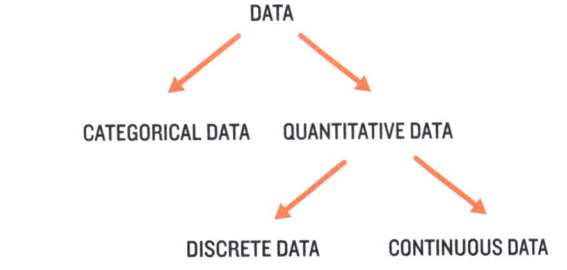

Figure 6: Data classification.

2.1.1 What Kind of Data Do I Have?

Say you have some general ideas about how to build your model and you think you have some data to employ. In order to properly analyze and make use of the data, the very first thing you must do is *classify the data,* as shown in Figure 6. The following questions can help.

Are the data words or categories (rather than quantities)?

Categorical data consist of names or labels, and we see this when the data can be divided into groups. For the NPS problem, in the NPS_hurricanes file you can see examples of categorical data. One column has the hurricane names, and another has the severity of the hurricane (which is also called a category!) such as ET, TD, TS, and H1–H5, as shown in Figure 7.

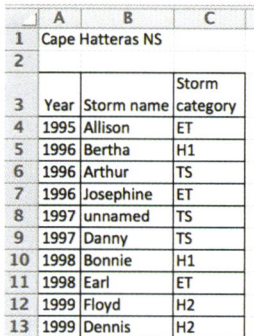

Figure 7: Hurricane names and classes are examples of categorical data. Data can be found at [2].

Are the data numerical quantities?

Data that express amounts are called *quantitative data*. Once you recognize that you have quantitative data, you need to classify it even further before you do any analysis.

- *Discrete data* consist of only a finite number of values (often integers). For example, the number of park rangers working at Acadia National Park each year would be discrete data (e.g., we'd never say there are 55.42 park rangers).
- *Continuous data* have an infinite number of possible values within a given range. The air temperature at Cape Hatteras National Seashore is considered continuous data (e.g., the temperature could take on a value of 55.42 degrees).

Hopefully you now feel confident with classification. There are a few other things to keep in mind when you are in the early stages of working with your data.

1. Some software has been trained to try to classify data on its own without any input from a user. This can be useful, because the software might automatically format the data nicely. Keep your eyes open, though, because, unfortunately, it may mis-recognize a data type and format things strangely.

2. It is possible that multiple data types are represented in a single data set, as we see in the `NPS_wildfires` data (see Figure 8). The column containing the cause of the wildfire has categories of "human" and "natural," whereas the year and the starting number of acres are quantitative.

	A	B	C	D	E	F	G
1	CalendarYear	FireName	IndexTime	SizeClass	CauseCateg	StartTime	StartAcres
2	1997	FASTCAR	6/28/1997 18:00	A	Human	6/28/1997 18:00	0.1
3	1997	KLEBERG1	7/2/1997 12:00	B	Human	7/2/1997 12:00	1
4	1998	DUNNRANCH	3/29/1998 16:00	E	Human	3/29/1998 16:00	0.1
5	1998	PANHANDLE	8/8/1998 12:55	D	Human	8/8/1998 12:55	5
6	1999	CAR	2/14/1999 12:30	B	Human	2/14/1999 12:30	0.5
7	1999	GREENGIANT	5/1/1999 17:30	C	Human	5/1/1999 17:30	0.7
8	2001	501	5/21/2001 21:00	B	Natural	5/21/2001 21:00	0.1
9	2001	FIRE 1011	10/10/2001 22:00	B	Natural	10/10/2001 22:00	0.1
10	2004	COLLINS FI	7/25/2004 2:30	E		7/25/2004 2:30	0.1
11	2004	BIG POND	12/24/2004 1:45	F	Human	12/24/2004 1:45	0.1
12	2005	HQ FIRE	1/16/2005 9:02	D	Human	1/16/2005 9:02	0.1

Figure 8: The `NPS_wildfires` data [2] shows how it's possible for a single data set to have both categorical and quantitative data.

3. Sometimes you can make a sensible translation between categorical and quantitative data. For example:

- Storm categories are based on wind speeds (for example, the NPS_hurricanes data [2] shows that a category 2 hurricane (H2) has wind speeds from 96 to 110 mph, as shown in Figure 9), so you could reasonably replace the categorical data with a corresponding wind speed or range of speeds.

- The NPS_wildfires data has categorical classes of fire, A through G (as shown in Figure 10), which could be translated to the quantities 1 through 7 for visualization purposes. *Use caution when doing this,* though, because sometimes one class might be many times more severe than another, as with the Richter scale for earthquakes, which is logarithmic.

In summary, if you do not know what kind of data you have, you will not be able to use it, share it, or explain it to other people, so it's important to always begin by classifying.

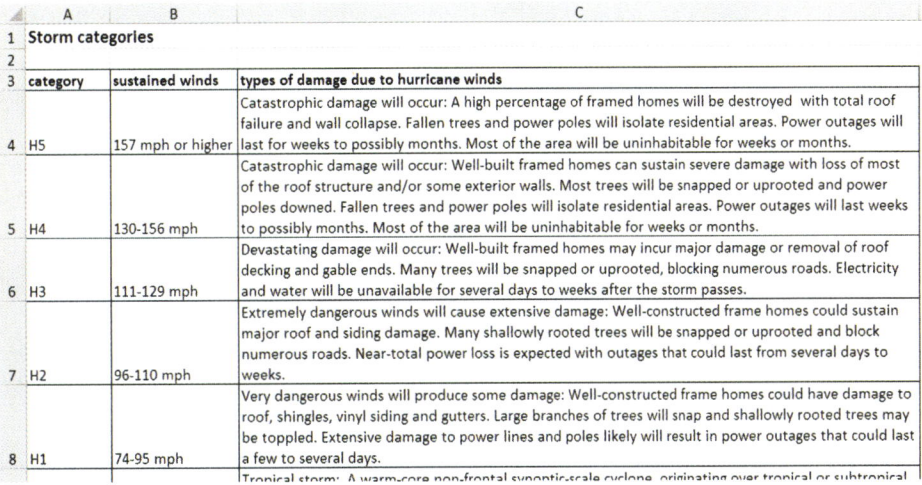

	A	B	C
1	Storm categories		
2			
3	category	sustained winds	types of damage due to hurricane winds
4	H5	157 mph or higher	Catastrophic damage will occur: A high percentage of framed homes will be destroyed with total roof failure and wall collapse. Fallen trees and power poles will isolate residential areas. Power outages will last for weeks to possibly months. Most of the area will be uninhabitable for weeks or months.
5	H4	130-156 mph	Catastrophic damage will occur: Well-built framed homes can sustain severe damage with loss of most of the roof structure and/or some exterior walls. Most trees will be snapped or uprooted and power poles downed. Fallen trees and power poles will isolate residential areas. Power outages will last weeks to possibly months. Most of the area will be uninhabitable for weeks or months.
6	H3	111-129 mph	Devastating damage will occur: Well-built framed homes may incur major damage or removal of roof decking and gable ends. Many trees will be snapped or uprooted, blocking numerous roads. Electricity and water will be unavailable for several days to weeks after the storm passes.
7	H2	96-110 mph	Extremely dangerous winds will cause extensive damage: Well-constructed frame homes could sustain major roof and siding damage. Many shallowly rooted trees will be snapped or uprooted and block numerous roads. Near-total power loss is expected with outages that could last from several days to weeks.
8	H1	74-95 mph	Very dangerous winds will produce some damage: Well-constructed frame homes could have damage to roof, shingles, vinyl siding and gutters. Large branches of trees will snap and shallowly rooted trees may be toppled. Extensive damage to power lines and poles likely will result in power outages that could last a few to several days.
			Tropical storm: A warm-core non-frontal synoptic-scale cyclone, originating over tropical or subtropical

Figure 9: The NPS_hurricane data for storm categories shows how it might be possible to make a sensible translation between categorical and quantitative data.

	A	B	C	D	E
1	Size Class Of Fire				
2	Class A - one-fourth acre or less;				
3	Class B - more than one-fourth acre, but less than 10 acres;				
4	Class C - 10 acres or more, but less than 100 acres;				
5	Class D - 100 acres or more, but less than 300 acres;				
6	Class E - 300 acres or more, but less than 1,000 acres;				
7	Class F - 1,000 acres or more, but less than 5,000 acres;				
8	Class G - 5,000 acres or more.				
9	https://www.nwcg.gov/term/glossary/size-class-of-fire%C2%A0				
10					

Figure 10: The NPS_wildfires data file gives quantitative descriptions of wildfire classes.

2.1.2 Are there any missing data or obviously wrong data?

A reality of data is that people (and sensors) make mistakes as they record it, or even fail to record it. Sometimes the mistakes are obvious, such as a negative value for someone's height. Other times the problems are harder (or impossible) to detect. Some basic things to check:

- Are there misspellings in categorical data?
- Do the numbers have the correct sign?
- Are the numbers within the expected range?

The process of examining data and addressing any missing or incorrect values is called *cleaning the data*. When some data values are missing or seem wrong, you have to make a modeling decision: Is the data set still worth using? Is there a reasonable way to fill in the missing values or discard suspicious/incorrect ones? Whatever you do, you'll need to document and justify your choices.

One common issue you need to be aware of is how your software will handle the situation where you are missing a data value. The software might

- give you an error,
- leave the blank and ignore it,
- assign a value of zero, or
- assign a nonsense value.

Missing data can be a big problem if the software assigns a value that affects your results and you didn't notice it was happening.

As an example, look at some of the `NPS_visitor_stats` data, as shown in Figure 11. The data includes some zeros. Upon closer inspection, you can guess that when the download was done in June, some of the May data had not yet been entered. One clue is that the previous month's value isn't zero, so you wouldn't expect a value of zero for the current month. You don't want the missing data to affect your analysis, so you might choose to use a row without zeros or go find another set of data.

Acadia NP
Report Date: May 2017

	This Month	Same Month Last Year	% Change	This Year YTD	Last Year YTD	% Change YTD	Fiscal YTD
Recreation Visitors							
Mount Desert Island	198,689	198,549	0.1	318,194	338,033	-5.9	723,397
Schoodic Peninsula	19,167	13,518	41.8	35,727	30,088	18.7	72,236
Isle au Haut	0	284	-100.0	0	284	-100.0	384
Visits from Island Explorer Bus System	0		0.0	0		0.0	0
Total Comm/Conc Bus Passengers	0	4,796	-100.0	0	4,796	-100.0	39,453
Total Recreation Visitors	217,856	217,147	0.3	353,921	373,201	-5.2	835,470
Non-recreation Visitors							
Mount Desert Island	6,000	6,000	0.0	6,000	6,000	0.0	10,500
Schoodic Peninsula	1,500	1,500	0.0	3,900	3,900	0.0	6,600
Total Non-Recreation Visitors	7,500	7,500	0.0	9,900	9,900	0.0	17,100

Figure 11: In this sample of National Park Visitor data [2], several pieces of data are missing. If a modeler wants to use this data, she'll have to make decisions about the appropriate way to handle the missing data.

Be aware that missing/incorrect data are not uncommon, and that you need to examine your data and account for it as you move forward in the modeling process.

2.1.3 Should I sort the data? How?

Sorting data is a useful way to take a first look and get a sense of what is there. It can help you quickly see if there are any negative values or any rows with missing data (since blanks will usually all end up together), and can help you identify outliers.

- Spreadsheet software: In spreadsheet software (for example, Excel) you usually select the data you want to sort by highlighting it, then look for a pull-down menu for data and sorting. Looking for sort in the help files will give you the details.

- MATLAB: The `sort` command has many ways to easily specify what values you want to sort.

- Statistical software R: You can sort a data frame in R using the `order( )` function. By default, sorting is ascending. To sort in descending order, put a minus sign in front of the sorting variable.

- Programming language (such as Python): In Python, the method `sort( )` sorts objects of a list.

If you're not quite sure how any of these commands work, you can do a quick internet search to find resources that can help. For example, if you're using MATLAB you might search on "MATLAB and sort".

2.1.4 Counting

A valuable early step in understanding your data might be to count values you identify as important to your model. Check out the following example to see why and how you might count.

Example: Counting efficiently

Consider the Cape Hatteras National Seashore hurricane data in Figure 12, which has three columns: Year, Storm name, and Storm category.

As you examine the data, the first thing you might notice is that there were no hurricanes in 2009 and 2010, while in some years, such as 1999, there were multiple storms. One year the storm was unnamed. This data set is relatively small, so you could probably count things by hand, but imagine you wanted to know the total number of category 1 hurricanes there were from 1995 to 2015.

In Excel, the command

$$=\texttt{COUNTIF(C4:C37,"=H1")}$$

gives the answer 4. Most programming environments have a similar function that can count items in a list that satisfy certain criteria.

If you want to know the total number of category 1 and category 2 hurricanes there were from

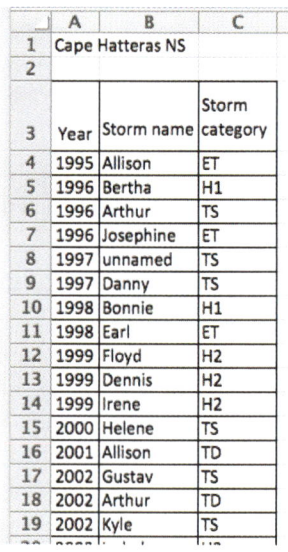

Figure 12: Partial view of the Cape Hatteras hurricane data [2].

1995 to 2015, you could do it this way:

$$\texttt{=COUNTIF(C4:C37,"H1") + COUNTIF(C4:C37,"H2")}.$$

To get the total number of all types of hurricanes H1–H5, you could use the "?", which is a wildcard, meaning that Excel will count anything that starts with an H, no matter what follows the H. The syntax looks like this:

$$\texttt{=COUNTIF(C4:C37,"=H?")}.$$

And to get the total number of hurricanes or tropical storms (but not ET) you can do this, where "<>" means "not equal to":

$$\texttt{=COUNTIF(C4:C37,"<>ET")}.$$

In a report you might want to include these counts as a way of summarizing your data for the reader (rather than showing it all). It could also help you answer questions such as "How often are most storms category 1?" that could give you a sense of how to construct your model.

2.2 VISUALIZE THE DATA

It is worth noting that, when you take a look at data or think about simulating it, you might consider the basic shape you expect the data to take. You may ask yourself: How are the data distributed? What kinds of shapes or trends can you see? Are there clusters? Is it symmetric or skewed? At this point, you may be realizing that visualization can be a powerful partner in using statistics, and you are right!

We have a whole chapter on visualization later, but here we are going to focus on a few ideas that relate to statistics. Your choices about visualization will depend on what type of data you have (categorical or quantitative) and how many variables you have. Take a look at Figure 13 to get some ideas. You probably recognize some of the suggested visuals, such as bar charts or scatterplots, but if not, no worries. It is easy to find examples online and even play with some software to see how they look. Most software comes with tutorials about how to plot data.

#Variables	Variable type	Suggested visual(s)
1	categorical	bar graph or pie chart
	quantitative	histogram or box plot
2	one categorical & one quantitative	line graph (e.g., categories over time)
2 (or more)	categorical	side-by-side bar graph
	quantitative	scatter plot

Figure 13: Relationship between variable type and visualization type. A more detailed discussion is provided in the Visualization chapter.

2.3 ANALYZING/ EXAMINING DATA

2.3.1 Finding the "middle" of your data

Mean, median, and *mode* are three kinds of averages, or ways of finding the middle of a numerical data set. Each uses a single number to describe an entire set of numbers.

- To find the mean, you divide the sum of all the items by the number of items.
- To find the median, you take the middle number, so that half the numbers are larger and half are smaller (if there are two middle numbers, you use their mean).
- The mode is the number that appears most often.

If your data are numerical, finding these measures of middle is fairly straightforward using technology.

- Spreadsheet software: In spreadsheet software the mean is sometimes called the average, whereas median and mode are called just that.
- MATLAB: MATLAB has `mean`, `median`, and `mode` functions.
- Statistical software R: The R language has `mean()` and `median()` functions but not mode.
- Programming language (such as Python): Python has `mean()`, `median()`, and `mode()` functions.

If you have continuous quantitative data, you can make a histogram of the data and look for the peaks. If instead you have categorical data, you can learn additional information about the variable by finding the mode. This can be achieved in Excel (see Figure 14) by:

1. identifying all categories,
2. using the `COUNTIF` function to find the number of occurrences of each storm type, and then
3. finding the largest number.

Technology can make finding averages/the middle of your data relatively easy to do, and captures some aspect of your data in a single number, but use this information carefully. Suppose, for example, you are modeling the total cost of several kinds of natural disasters. You

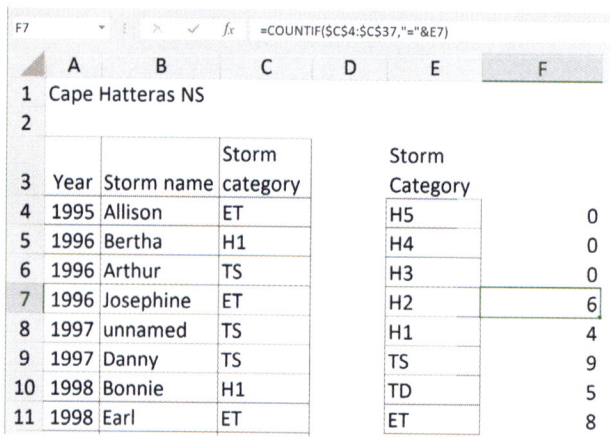

Figure 14: Storm severity at Cape Hatteras National Seashore [2].

want to incorporate the number of each kind of disaster you might expect in a year, given the data provided in the NPS problem. How could you use an average and the hurricane data to find a value for the number of storms you expect each year? Using this one number simplifies the model, but it means you won't be able to see the ways the number of storms varies from year to year (which might be important, depending on the goal of your model).

You could also run into trouble by averaging together numbers in a way that doesn't make sense. For example, if you average the year of a storm with the category of a storm, you can get a value, but it doesn't provide useful information. Why not? Say you average 2015 with a category 2 storm and 2013 with a category 4 storm. In both cases you get the number 1008.5, but that number doesn't give you any new insight. It is also questionable to add numbers that have different units.

2.3.2 Distributions/Trends/Clusters

In statistics, *distributions* can help you see not only the possible values in your data set, but how often they occur. You may have heard about the popular "bell-curve," which is officially called a *normal distribution*. The normal distribution looks like a symmetric bell and is determined by the mean and standard deviation of the data set. We won't go into the nitty-gritty detail here, but you may wish to step back and refresh your memory (or familiarize yourself) with the notion of a distribution and the difference between normal, uniform, and bi-modal distributions.

If your variable of interest is a single quantitative variable, you can learn a lot about the distribution of your data by producing a histogram. Because histograms represent the occurrence of a variable over a chosen interval, different visual features such as "clustering" can provide valuable insight into the nature of a data set. Let's consider an example with some clustering.

Suppose you are interested in determining a sea level change "risk rating" for locations in the United States; in particular, you want to identify what it means for a coastal region to be at "higher" or "lower" risk for sea level change. A histogram created using mean sea level (MSL) data from all tide stations in the National Water Level Observation Network (Figure 15) shows the frequency of occurrence of MSL rates at 142 different observation stations at locations across the United States as well as in a number of U.S. Territories.

In Figure 15, you can see that a good portion of the data is centralized or clustered between 0 and 6 mm/yr, and there is one peak (we call this unimodal). As a result, it may be reasonable to assume that MSL is normally distributed. Further investigation of the data (perhaps identified through a sort based on station location) reveals that some of the reporting stations are located in U.S. Territories as well as in Alaska. Removing these stations, and focusing instead on information from monitoring stations based solely in the continental U.S., returns a histogram that appears to be normally distributed (Figure 16). With this information, you can choose appropriate descriptive statistics to further analyze this situation.

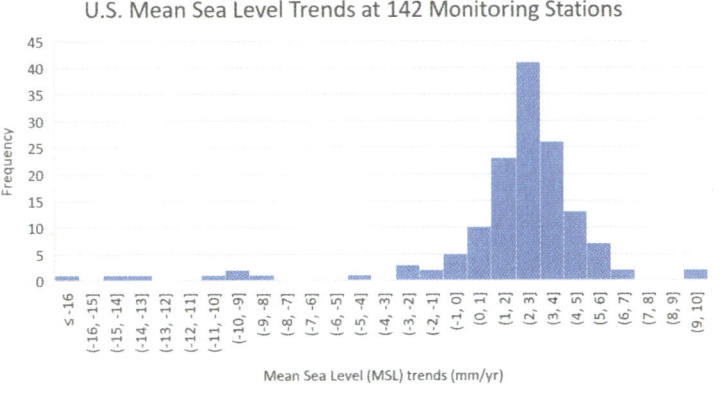

Figure 15: Mean sea level (MSL) trends calculated using MSL observations from 142 long-term (minimum 30 years of data) tidal stations located in the United States and U.S. Territories [8]. For example, there are 25 monitoring stations in the category (3, 4]. This means that at each of those 25 stations, the mean sea level went up between 3 and 4 mm per year.

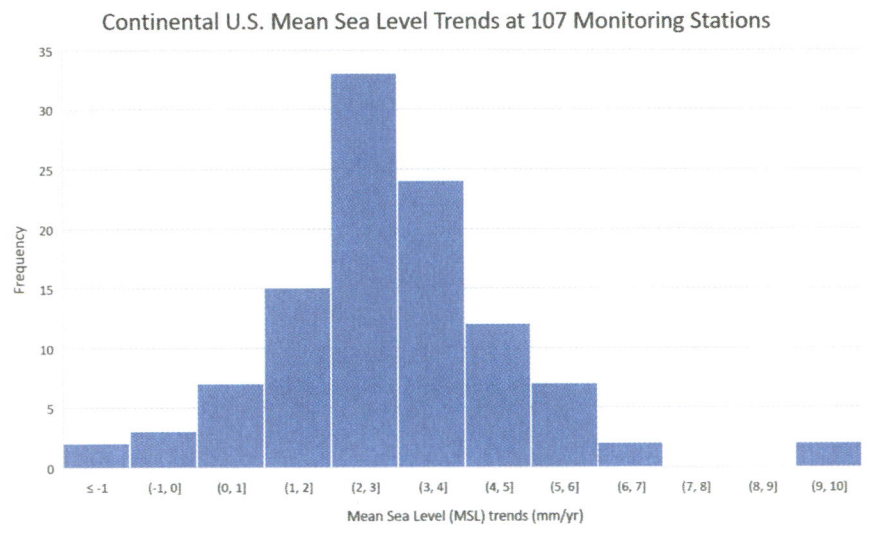

Figure 16: Mean sea level (MSL) trends calculated using MSL observations from 107 long-term (minimum 30 years of data) tidal stations located in the continental United States [8].

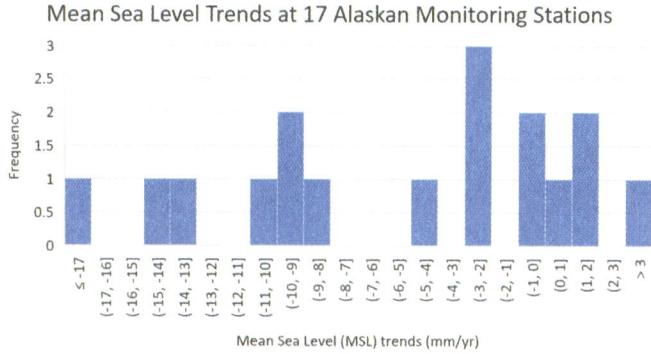

Figure 17: Mean sea level (MSL) trends calculated using MSL observations from 17 long-term (minimum 30 years of data) tidal stations located in Alaska [8].

It may be valuable to take a look at other collections of the data. For example, the majority of the candidates for outliers are values taken from observation stations off the coast of Alaska. Looking at those data points independently (Figure 17), we see that they are not distributed normally. This realization may allow us to develop a strategy for rating the risk in the Alaska data as well.

Another useful approach is to compute the mean and median values associated with your quantitative data (as demonstrated earlier in this section) and compare the two. If the numbers are fairly close in value, it implies symmetry in your data, which *could* imply a normal distribution. But be careful! It may also represent other symmetries such as a bimodal distribution or even a (nearly) uniform data set. In the case of the MSL trends, the Continental U.S. data visualized in Figure 16 has mean = 2.94 and median = 2.83; while the same data from Alaskan monitoring stations (Figure 17) has mean = -5.31 and median = -2.69. As a result, we expect the Continental U.S. data to be symmetrically distributed, while we don't expect the same for the Alaska data, which is consistent with our earlier findings.

2.3.3 Line/Curve fitting

When you believe you have data for more than one variable (say, height and calories eaten in a day) and you believe there is a relationship between the two variables, you can create a scatter plot of points and see if it indicates a relationship.

The reason this can be useful is that if there is a strong functional relationship between the variables, you can use the function to provide values that are missing or to see what might happen for values outside your given data set (extrapolation, a much more risky venture). The art and science of line/curve fitting is an important topic in statistics. We can't teach you all the details here, but we can point you to some useful tools.

What is r-squared?

After plotting data and noting the likelihood of a relationship between a variable of interest and one (or more) independent variables, you may want to run a *regression analysis* (see how below), which will provide one measure of the strength of the linear relationship.

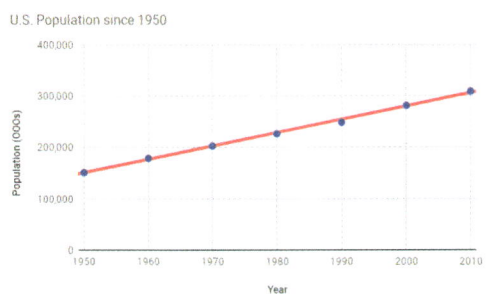

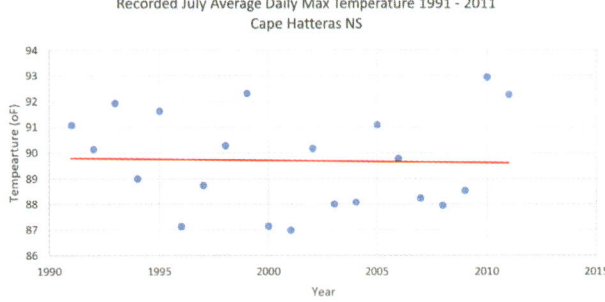

(a) This data [9] appears to be close to linear, which is confirmed by its r^2 value ($r^2 = 0.997$), which is close to 1.

(b) This data set [2] does not exhibit an obvious linear trend, and this is reflected in the r^2 value ($r^2 = 0.0009$), which is close to 0.

Figure 18: Best-fit lines through two different data sets.

If you choose to include a function that was determined via regression in your modeling solution, you need to justify why it's reasonable to use this modeling approach to estimate values of your variable of interest. One of the most common (and frequently misused) ways to do this is to report a value for r-squared (r^2) associated with a regression performed on a set of data. The r^2 value is defined to be the ratio of the variation in the modeled results to the variation in the actual data set and measures how close actual data is to those that were predicted by a regression curve.

Because r^2 is a ratio with a denominator that is always larger than the numerator, it can only take values between 0 and 1. A value of $r^2 = 1$ suggests a perfect correlation; $r^2 = 0$ indicates that the model isn't providing any insight into the variation in the original data. It is important to note that finding a value for r^2 rarely tells the entire "story." For example, a linear regression with $r^2 = 0$ means that there is not evidence of a *linear* relationship between the variables; there could still be a nonlinear relationship, or another less apparent connection between the variables. In Figure 18 you can see examples of data with linear fits, one with an r^2 value close to 1, and one with an r^2 value close to 0.[3]

When you communicate your regression in a report or presentation, it's important that you not only report the r^2 value, but that you also explain what the r^2 value means, so your audience will have more confidence in your model. Many people see r^2 close to 1 as justification for using the regression they found in future calculations because this means your model fits the data pretty well. Despite this general guideline, you should not rely on r^2 alone to justify your decision. We'll explain this in an example later.

Mechanics of curve-fitting

You can use all sorts of technology to perform a line/curve fit and calculate r^2:

- Spreadsheet software: Spreadsheet software usually has options for linear, polynomial, exponential, and power law fits. For a polynomial fit, choose an appropriate degree (see guidance below). You may want to do an internet search on "linear regression" to learn how, and whether you would like to display the associated r^2 value.

- MATLAB: You can get started with `polyfit` and `polyval`, but in order to determine whether the fit is appropriate, you may need to do an Internet search on "regression in MATLAB" to see examples of finding the associated r^2 value. For example, you can plot the data and then find Basic Fitting under the Tools tab in the figure window. You also have the option of purchasing the Curve Fitting Toolbox, which has built-in GUI for fitting curves and surfaces to data. With the Curve Fitting Toolbox, `cftool` will give you the r^2 value.

- Statistical software R: After identifying your variables, you have a choice of functions to call on based on the type of regression you plan to run. For example, using `lm()` will perform a least-squares regression; a subsequent `summary()` command will return details on the fit, including r^2 and adjusted r^2.

- Programming language (such as Python): Commands `polyfit` and `polyval` will work for linear regression. As with MATLAB, an Internet search will help you determine the correct syntax for finding an r^2 value.

Is a line the best?

For any data set you have, you can use technology to find the best-fit line, but is a line the best fit for the data? Would an exponential function be better? How can we know?

From a modeling perspective, the simplest reasonable fit is usually the best. Many modelers start with a linear fit and can adjust their model later if it makes sense. A linear fit is appropriate when the rate of change appears to be constant. If it looks like the rate of change is increasing, you might try a higher degree polynomial, a power law, or an exponential fit; if the rate of change decreases, you might try a logarithm or hyperbolic tangent. Making a plot of the data to see what it looks like is the most important first step.

The r^2 value can provide some guidance to help you know whether the curve is a good fit for your data. However, the r^2 value does not tell the entire story! Look at the following examples that demonstrate this and show some of the pitfalls you'll want to avoid.

Example: Using residuals.

To demonstrate a point about the r^2 value, let's look at two data sets: monthly MSL information collected from Cape May, NJ since 1978 [7], and carbon dioxide measurements from Mauna Loa Observatory [6]. Scatter plots of both sets of data are presented in Figure 19. In both data sets there is a single independent variable and a single response variable. For the MSL data (Figure 19a), the red line represents the line of best fit found by using linear regression to model the relationship between time (in months) and monthly MSL since September, 1978; r^2 = 0.516. Similarly, the red line associated with the carbon dioxide measurements at Mauna Loa Observatory (Figure 19b) shows the line of best fit that models the relationship between time (in years) and atmospheric carbon dioxide observed at Mauna Loa since 1959; r^2 = 0.985.

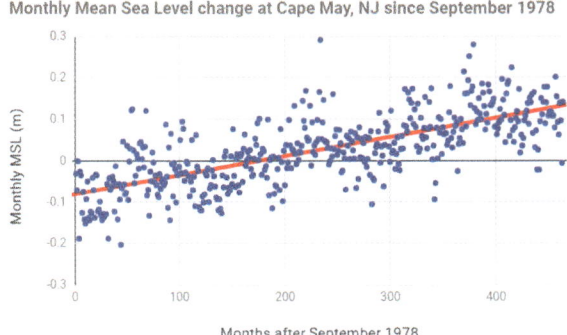

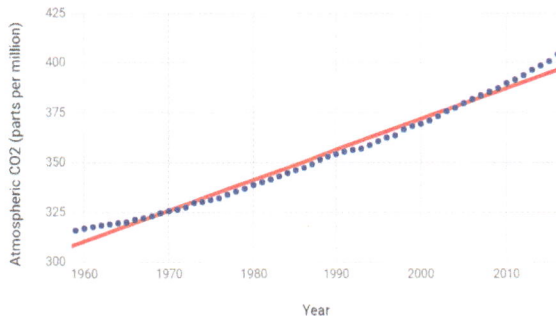

(a) Linear regression associated with monthly MSL at Cape May, NJ [7] has an r^2 value of 0.516.

(b) Linear regression associated with carbon dioxide measurements at Mauna Loa Observatory [6] has an r^2 value of 0.985

Figure 19: Linear regression for two different data sets.

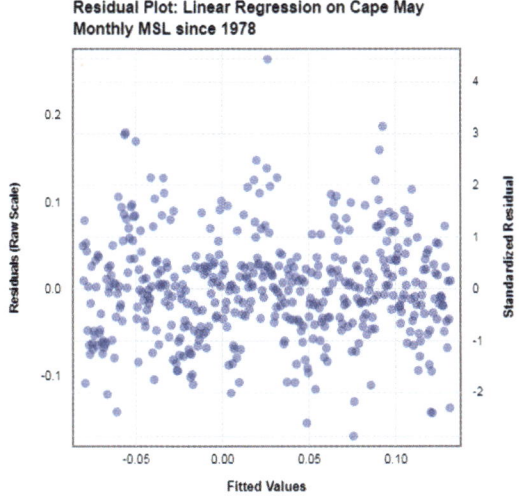

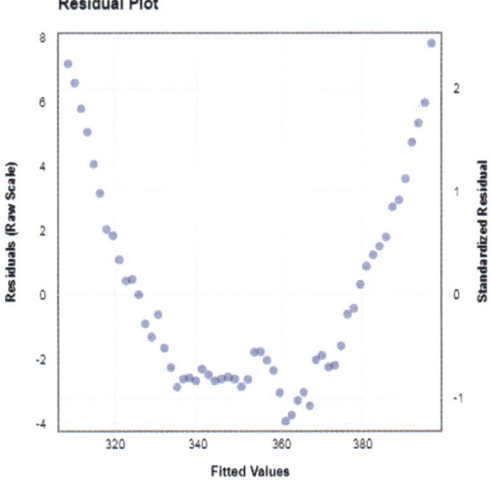

(a) The residuals for a linear fit of monthly MSL at Cape May, NJ [7].

(b) The residuals for a linear fit of carbon dioxide measurements at Mauna Loa Observatory [6].

Figure 20: Residuals for linear regressions of two different data sets.

If we try to determine which of these sets of data is best represented by a linear model, and we look only at their r^2 values, we would probably assume that the fit shown in Figure 19a is inferior. One way we can further investigate is by looking at a plot of the residuals. A residual is the difference between the observed value (i.e., the data) and the predicted value (i.e., the value of the model). Figure 20 shows the residuals for the linear fits we found in Figure 19.

Despite a significantly higher r^2 value for the linear model associated with carbon dioxide measurements at Mauna Loa over time, it may be a mistake to use the linear model there, because the residuals (shown in Figure 20b) are clearly not randomly distributed. If the plot of residuals is not randomly distributed and exhibits a distinct pattern, this means we *may* be missing some key feature of the data that the model does not capture. Quite the opposite could be said for the residuals associated with Cape May's Monthly MSL, Figure 20a; there is no clear or easily identifiable pattern, and approximately half of the residuals are positive while half are negative. As a result, it may be reasonable to use a linear regression model to make some predictions for future MSL in Cape May (even though the r^2 value is not close to 1).

Since we may wish to improve upon the linear model we initially made for the carbon dioxide data, we can try other approaches. We might consider an exponential fit (see Figure 21), but it turns out that this data is well fit with a quadratic function, as evidenced by the plots of the data, best-fit quadratic function, and corresponding residuals in Figure 22. The r^2 value is close to 1, and the plot of the residuals appears to be randomly distributed.

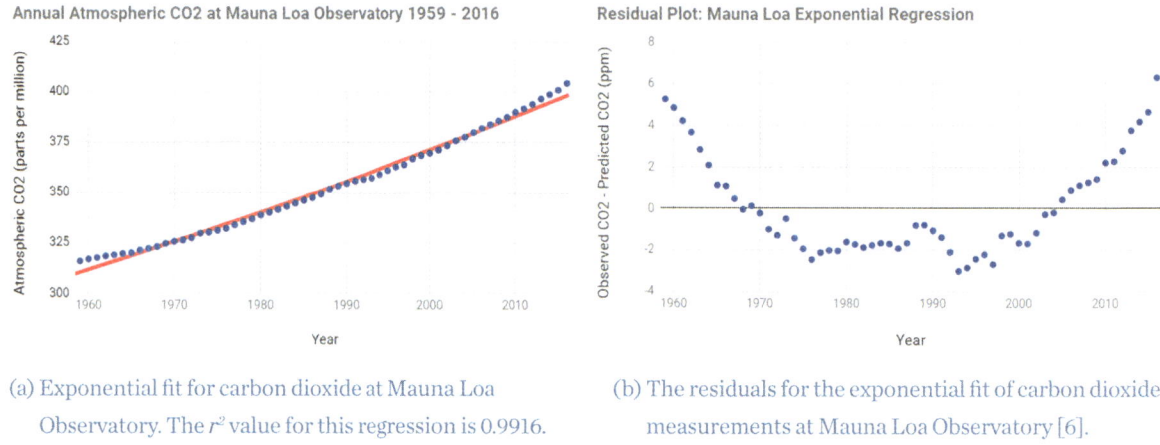

(a) Exponential fit for carbon dioxide at Mauna Loa Observatory. The r^2 value for this regression is 0.9916.

(b) The residuals for the exponential fit of carbon dioxide measurements at Mauna Loa Observatory [6].

Figure 21: Exponential fit and corresponding residuals for carbon dioxide at Mauna Loa Observatory [6].

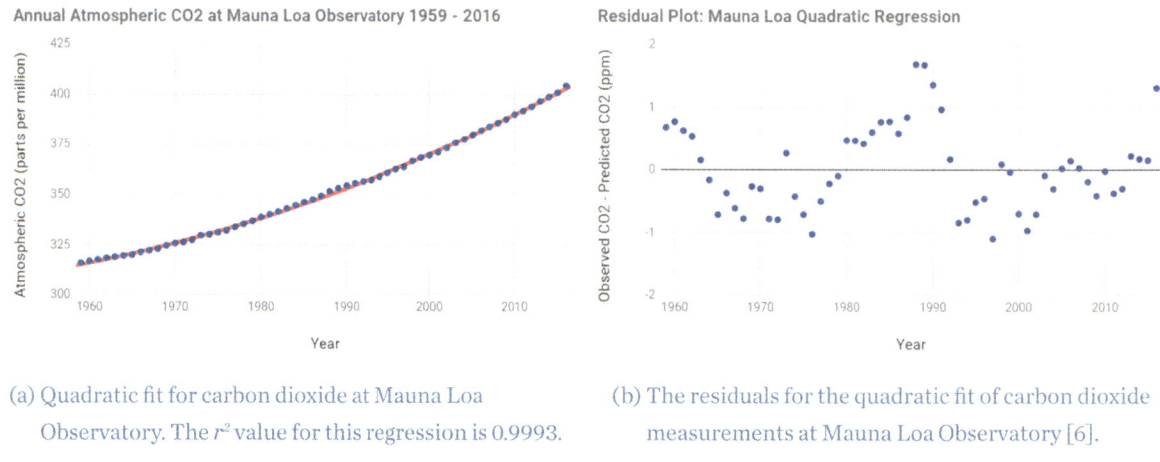

(a) Quadratic fit for carbon dioxide at Mauna Loa Observatory. The r^2 value for this regression is 0.9993.

(b) The residuals for the quadratic fit of carbon dioxide measurements at Mauna Loa Observatory [6].

Figure 22: Quadratic fit and corresponding residuals for carbon dioxide at Mauna Loa Observatory [6].

The key idea here is that once you've performed a regression and determined the r^2 value for your fit, it's worth making a plot of the residuals. If the residuals aren't randomly distributed, then you should think about whether your model is really appropriate.

It's worth noting just one more aspect of the carbon dioxide data, and that is that the r^2 value for the linear fit is 0.985, which is one sign that a linear fit may be appropriate. The plot of the residuals associated with the linear fit (Figure 20b) is not randomly distributed, but the residuals themselves are all between –8 and 4. When we consider that the data lies between 300 and 425 (as shown in Figure 20b), we can conclude that the error in using the linear model is less than 5%. If you were trying to estimate missing values in between those given in the data, the linear fit might be sufficient. Since linear models are simple to use and explain, it might actually make sense to use the linear model (even though the quadratic model might be "better" in some ways). On the other hand, if you wanted to use this model to estimate values in the future (also called extrapolation), you might have greater confidence in the quadratic model. You as the modeler have choices about which fit you might continue with under these circumstances. You'll want to communicate how you made your decision and what information you used to make it.

At this point you should be convinced that the r^2 value could be misleading and that residuals can be useful in making decisions about curve-fitting. There's one more mistake you should avoid, and we'll discuss it below.

Example: Residuals and r^2...still not enough!

Consider the data given in Table 1 for the average daily maximum heat index for the month of October.

Year	Avg. daily max heat index (in degrees F)
1997	82.72
2002	86.48
2007	86.14
2008	82.4
2009	85.74

Table 1: Average daily maximum heat index for the month of October from 1997–2009 in Cape Hatteras, NC [2].

Notice that there is a lot of missing data! As a modeler, you might want to consider investigating a different month or perhaps look for a different data source. Let's assume you've decided to move forward with this data set and you'd like to see what kind of fit makes sense.

Spreadsheet software often makes it really easy to quickly show multiple types of regression,

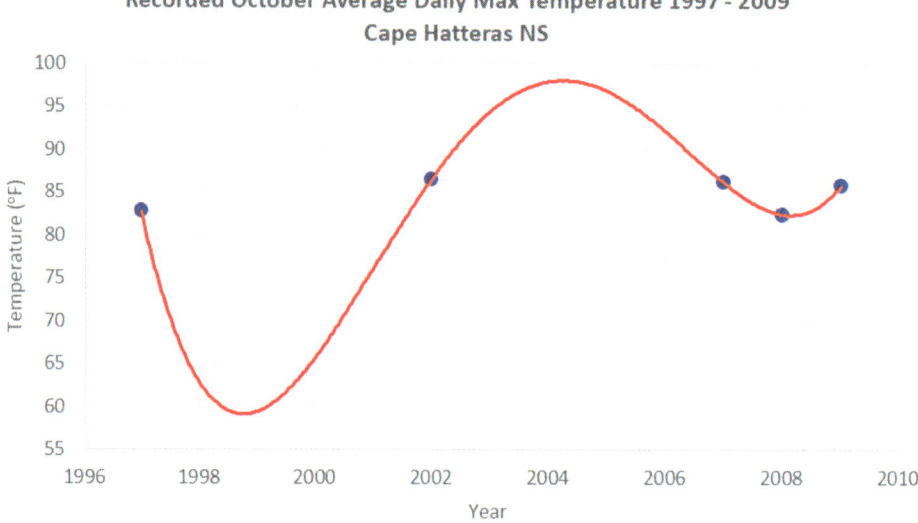

Figure 23: Average daily maximum heat index for the month of October from 1997-2009 in Cape Hatteras, NC [2], and the best-fit fourth-degree polynomial ($r^2 = 1$).

so that you can quickly look at plots and r^2 values and determine the right model for the given situation. In this case, you might see what happens if you decide to fit the data with a fourth-degree polynomial. Figure 23 shows the plot of the data and the best-fit fourth-degree polynomial.

At first glance, this might seem like a great choice for a regression. It looks like the curve goes through every single one of the data points perfectly, and we can see that it actually does because the r^2 value is exactly 1!

Imagine that we want to use this fourth-degree polynomial model to estimate some of the missing data, say for the year 1990. We can see in Figure 23 that the estimate would be approximately 60 degrees. This seems inappropriate given that the nearest data are 82.72 and 86.48 (in 1997 and 2002, respectively). Wouldn't it make more sense to estimate a value between 82 and 87? Further, can you envision what the model would estimate for year 1996? Looks like it would be 100 or more, which is totally unrealistic. What's going on here?

Notice that our data set consists of only five points, and we are trying to find a fourth-degree polynomial, which has the general form

$$y = ax^4 + bx^3 + cx^2 + dx + e.$$

In other words, we have used technology to help us find the coefficients $a, b, c, d,$ and e which will make the polynomial best represent the data. We have five data points and five unknowns, which gives us a well-defined system of equations. This means that the polynomial is guaranteed to go through all of those data points.

In general, when you use a polynomial of degree N−1 to model a data set with N points, your polynomial will go through every point, and you will find that $r^2 = 1$. That is, you can always fit a polynomial perfectly by making it of high enough degree, but that doesn't mean you have the best model. This is called overfitting and almost always results in a worthless model which cannot be trusted for estimating values.

For this data set, a linear regression (shown in Figure 24) yields an r^2 value of 0.0719, which is much closer to 0 than to 1. Nonetheless, if we had only these five data points, we had no other data available, and we absolutely needed to make estimates, the linear fit would be more reasonable in this instance. Remember the principle that the simplest fit is usually the best fit.

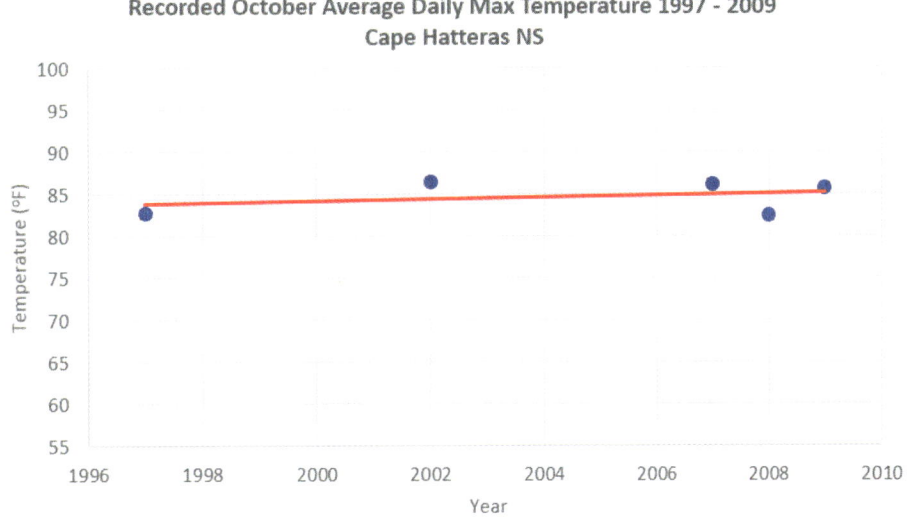

Figure 24: Average daily maximum heat index for the month of October from 1997-2009 in Cape Hatteras, NC [2], and the best-fit line (r^2= 0.0719).

Final thoughts on line/curve fitting

In this section, we touched upon some important concepts to keep in mind as you perform line/curve fitting, but you must always ask yourself as a modeler if what you're doing is reasonable.

Important considerations while curve fitting:

- If you plot your data and fitted curve together, does the result look sensible?
- Have you explained how you decided to choose the line/curve you chose?
- Have you communicated that you know what r^2 signifies?
- Have you plotted residuals to see if they exhibit a pattern?
- Does your regression-based model make sense in the real world?
- If you're using your model to extrapolate information, have you justified why the trend is likely to continue in the future?
- Are you overfitting with a high-degree polynomial?

Of course, we have just scratched the surface of regression analysis, and while we are confident that using this information can help you get started, we also encourage those of you with additional questions to take a statistics course and/or seek out additional resources to fully investigate the details of curve fitting using regression.

2.3.4 Extreme/Unusual data

Plotting data allows you to quickly locate possible outliers, see data inconsistencies, and identify trends which may help in building your model (consider heading over to the Visualization chapter if you need help plotting). Unusual values might indicate errors in data entry, or they could represent real but unusual instances.

As mentioned earlier, if your data set contains two quantitative variables, you should consider plotting the data pairs. Data points that you identify as *outliers,* that is, points that stand out in a data set, should not be removed immediately; in fact, taking time to investigate why a subset of data appears to be disconnected from the larger collection may be beneficial to the development of your model.

For example, you may be interested in forecasting how many visitors will travel to Acadia National Park in the near future. After accessing annual visitor data you plot the data (see Figure 25) and notice a clear break in the data set that may affect how you choose to use the data moving forward.

If you choose to use the full data set in your calculations, you might observe an overall (very weak, but nonetheless) negative correlation between time and number of visitors. However, a quick internet search (e.g., "Acadia National Park 1989 visitors") returns a collection of news stories that note that Acadia National Park changed the methodology used to determine visitors to the park in 1990. One writes[4]:

> "Because of differences in the complex array of statistics and formulas used in calculating visitation, park officials avoid comparing pre-1990 visitation estimates with those compiled since 1990."

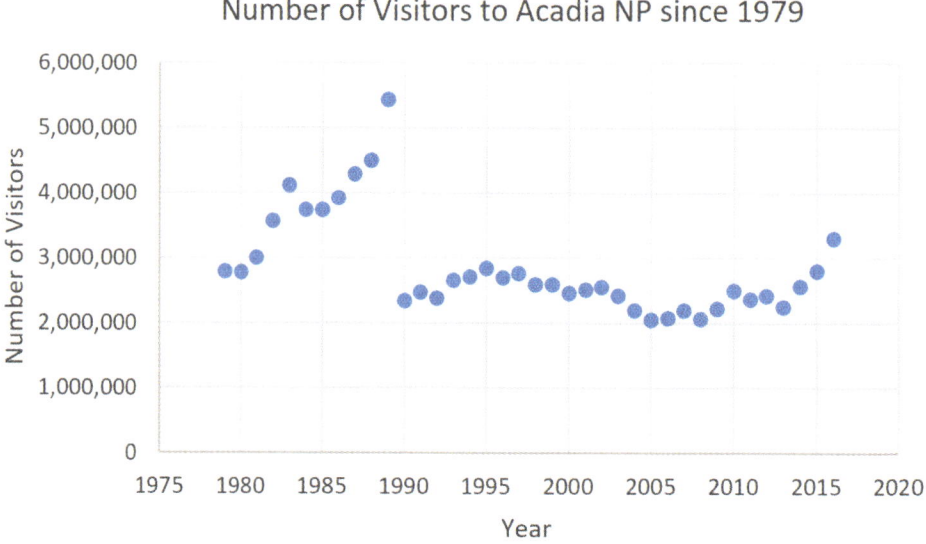

Figure 25: Excel plot of Acadia visitor data [2].

Further reading of the same article provides additional information:

> "Acadia officials have said they have confidence in the general accuracy of the year-to-year trends in the pre-1990 figures."

As a result, you may find it worthwhile to use all the data or just a portion of it (say, only before or after the 1990 change) in your model. Regardless of what data you choose to use, you need to share the reason you are making your choice with your audience.

The bottom line is that you should examine your data to identify any extreme/unusual data, investigate to see whether you can explain the phenomenon, and make sure you communicate about the choices you make as you proceed with your model.

2.3.5 Is this data actually useful to me?

The data set chosen might not work for you—if not, adapt what you are doing or find some other data to work with. The great thing about modeling is that you are usually free to use any data resources you want. This might lead you to wonder whether you should (a) create a model based on the data you have available or (b) create a model and then find (or simulate) data. In a modeling competition or with a short deadline, (a) might be a more efficient choice, but you still need to be selective so that your model isn't overly complicated by trying to use all the data that are provided.

2.4 TAKE AWAY MESSAGES

The real world is continuous, but in order to develop understanding, we often need to discretize. Statistics is a powerful modeling tool. You can use concepts from statistics effectively if you keep these ideas in mind:

- As you download or simulate data, you'll need to decide what type(s) of data you have and how they will work in your model.
- When you use software, some programs will make assumptions about what category or data type it should assign to the data you upload. It may or may not do this correctly.
- Sometimes it may be useful to make a sensible translation between categorical and numerical data types.
- You might also consider how the types of data you use will relate to the visualizations you create to communicate your analysis or results.
- Even when the data file is extremely big, you can look at sections or samples of it or count parts of it to get a handle on what is there.
- Regression is a common tool for analyzing data, but be careful because you can have a poor fit in several different ways. The r^2 value alone is not enough to justify your fit.

Concluding thoughts on statistics

Technology is a powerful resource for working with data. Statistical software and spreadsheets facilitate the use of data in models and allow for quick computation of descriptive statistics. As you decide which data to use and how to analyze it, documenting your reasoning for your choices will make your solution more understandable (and, if the choices are sound, more trustworthy) for your reader. In addition, visualization can play a key role in your statistical analysis and in communicating your results. We describe this last step more in the next chapter.

ENDNOTES

[3] Similar arguments can be made for what is called the correlation coefficient r. We use r^2 because it can be applied to models with more variables.

[4] B. TROTTER, Bangor Daily News: Acadia smashes park visitation record in 2016, January 18, 2017, *https://bangordailynews.com/2017/01/18/news/hancock/acadia-smashes-park-visitation-record-in-2016/*.

VISUALIZATION

Visualization is a key tool in modeling and can be useful at any phase. Specifically, you may choose to create a plot, graph, or chart in order to

- summarize data,
- make comparisons,
- reveal patterns (i.e., correlations, relationships),
- add interest to a report or presentation,
- demonstrate points, and/or
- facilitate understanding of information.

Visualization in the context of modeling is deeper than just plotting a function or creating a scatter plot. There is real-world significance in the underlying problem you are trying to understand, and visualization can help to provide insight. We consider two distinct reasons you may want to create a visualization during the mathematical modeling process.

As you are in the process of developing a model and finding results, visualizations can help you understand data you have found and/or can help you interpret the output of your model. We will call these uses *internal uses,* because they are mainly made to help you move forward in the modeling process. The visualizations you produce for internal use need not be fancy; they simply need to help you make decisions (so they need to be useful and correct).

Visualizations also help you convey your ideas to others, and we will discuss this in the section on *external uses.* Visualizations created for external use are extremely important; often when reading a text book (or technical report), our eyes go straight to the graphics to try to get the big picture or main idea. Since your audience will undoubtedly be drawn to your graphics, it's important to think about whether the visualization will make sense to someone who is unfamiliar with your modeling question. We'll discuss how you can make your visualization most understandable to a wide audience so you can communicate results and demonstrate the power of modeling!

There are many ways to display numerical results or data, so choosing the right way can be the difference between a reader understanding the key point or being totally confused. Some examples are scatter plots, line graphs, bar graphs, or pie charts. Even a flow chart or a drawing can be used to describe a complex process that would otherwise require an overwhelming amount of text to explain. On the other hand, sometimes a table is a straightforward way to show results. The third section in this chapter discusses several examples that demonstrate choices made for both internal and external visualization.

x1	y1		x2	y2		x3	y3		x4	y4
10.000	8.0400		10.000	9.1400		10.000	7.4600		8.000	6.5800
8.0000	6.9500		8.0000	8.1400		8.0000	6.7700		8.0000	5.7600
13.0000	7.5800		13.0000	8.7400		13.0000	12.7400		8.0000	7.7100
9.0000	8.8100		9.0000	8.7700		9.0000	7.1100		8.0000	8.8400
11.0000	8.3300		11.0000	9.2600		11.0000	7.8100		8.0000	8.4700
14.0000	9.9600		14.0000	8.1000		14.0000	8.8400		8.0000	7.0400
6.0000	7.2400		6.0000	6.1300		6.0000	6.0800		8.0000	5.2500
4.0000	4.2600		4.0000	3.1000		4.0000	5.3900		19.0000	12.500
12.0000	10.8400		12.0000	9.1300		12.0000	8.1500		8.0000	5.5600
7.0000	4.8200		7.0000	7.2600		7.0000	6.4200		8.0000	7.9100
5.0000	5.6800		5.0000	4.7400		5.0000	5.7300		8.0000	6.8900

Table 2: *Anscombe's quartet. The four data sets shown have nearly identical mean x-values, mean y-values, and variances in x and in y.*

3.1.1 Why should I visualize?

When you obtain a data set, descriptive statistics like the ones we discussed in the Statistics chapter are useful, but they may not reveal the underlying relationship between the variables. A classical example of why you should visualize data is demonstrated in *Anscombe's quartet* [10], a set of four data sets, shown in Table 2, given as x and y values, such that for each of the data sets the following hold:

• the mean of the x-values is 9,
• the variance in the x-values is 10,
• the mean of the y-values is 7.50 (to two decimal places),
• the variance in the y-values is 3.75 (to two decimal places),
• the line of best fit is $y = 3.00x + 0.500$ (to two decimal places), and
• the r^2 value for the linear fit is 0.816 (to three decimal places).

Based on this quick statistical analysis, you would likely believe that these data sets are really similar. Let's see what happens, however, when we plot the data, as in Figure 26. (Note: We created these graphics using MATLAB's plot and scatter tools.)

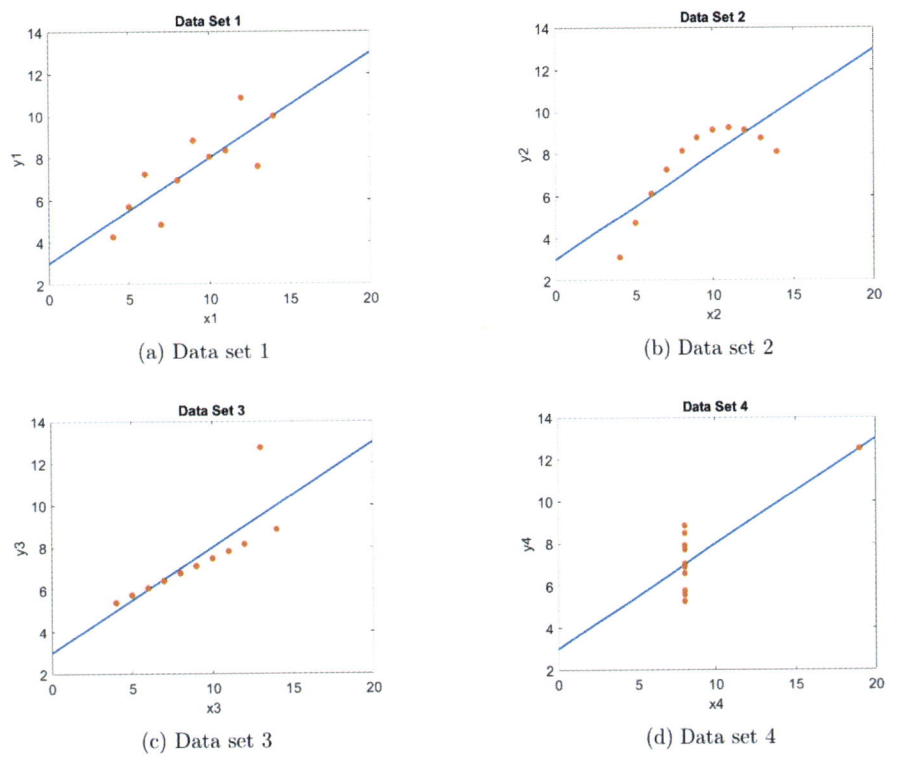

Figure 26: *Anscombe's quartet.* Data and lines of best fit are shown. Each of these data sets has the same summary statistics.

Each plot has a remarkably different shape from those of the other three. If you had started by just looking at the plots, you probably would not have guessed that their summary statistics are nearly identical. The moral of the story here is that the statistical analysis, although useful, doesn't tell the whole story about the relationship between the x and y values. ==Visualization of data is an important step in learning about data and should not be overlooked.==

3.1.2 How do I create a plot?

Spreadsheets, software tools, and programming languages are all equipped with a wide range of visualization tools that can help with your internal investigation (and external, for that matter).

One situation that arises frequently in modeling is when you have a set of points (x, y), which are both numerical data. Because this type of data is so common, we'll describe the process of creating a plot for such data, which may help uncover a relationship between x and y. Here are some examples of how software can be used, but this should not be considered an exhaustive list!

- Spreadsheet software: In spreadsheet software you usually select the data you want to plot by highlighting it, then find Insert on the command ribbon and choose your preferred scatter chart or plot format.
- MATLAB: Use the `scatter` or `plot` commands.
- Statistical software R: Use the `plot` command. You may need to pair up values to create (x, y) coordinates using `cbind`.

• Making a plot using a programming language, such as Python, may require installation of additional tools. When you read about programming languages in the Resources chapter, we mention something called a"library," which can be used within a programming language to give access to those tools. In Python, the library Matplotlib is used for visualization, including plotting.

While not every plot, graph, or figure you generate will appear in your final report, you should keep everything you generate just in case you do need it later. If possible, you should always aim to save graphics, especially those you expect to share, as independent files that have meaningful names for when you go back to find them. Be aware that some file formats can be edited easily later (like MATLAB .fig files), while others (like image files) could have to be remade from scratch. More on this below.

3.1.3 What type of visualization should I use?

Is your data classified and/or organized?

Suppose that you have obtained a collection of data points in an accessible (to you) software setting (e.g., spreadsheet program). There are MANY types of visualizations to choose from, so how can you know which is the appropriate one for your data? Before you can determine the best way to visualize your data, you'll need to classify, and possibly sort, the data. Recall that data may be either categorical or quantitative, as shown in Figure 27.

We discuss classifying and preparing data in more detail in section 2.1, so you may find it helpful to look in that section if the vocabulary that we use here is unfamiliar to you.

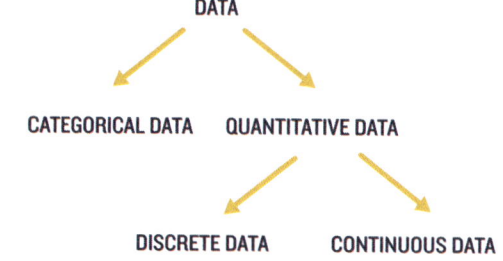

Figure 27: Data classification.

Choosing a visualization

Now that you've prepared your data, you can use the guidelines shown in Figure 28 to help you determine which type of visualization tool might be most useful.

You'll notice that each case appearing in Figure 28 names several options for potential visuals, and Table 3 provides considerations for making each type of visual.

Note that Table 3 simply provides guidelines; there are no "rules" that dictate exactly which form of visualization is the single correct one. Good visualizations should help you understand

#Variables	Variable type	Suggested visual(s)
1	categorical	bar graph or pie chart
	quantitative	histogram or box plot
2	one categorical & one quantitative	line graph (e.g., categories over time)
2 (or more)	categorical	side-by-side bar graph
	quantitative	scatter plot

Figure 28: Relationship between variable type and visualization type.

your data and perhaps help you make decisions about what to do next in the modeling process, as shown in Figure 29.

The examples at the end of this section describe how you might make choices about visualization based on the data you have. Here we focus on advice for polishing up visualizations before they go into a report or presentation.

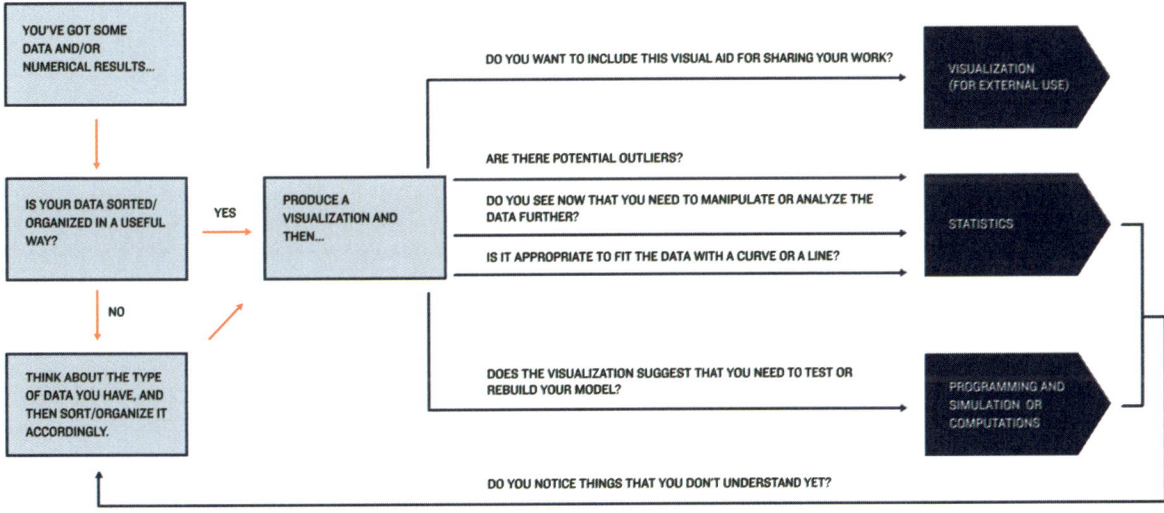

Figure 29: Decision flow chart to illustrate how questions may arise when using visualization for internal use.

Visualization Type	Best Uses	Considerations/Tips	Example
Pie chart	• One data set only • Data is parts of a whole	• Use contrasting colors • Use only when there are 7 or fewer categories • Won't work if any data are negative	
Bar graph	A mix of categorical and numerical data	• Used to compare sets of data between different groups • Bar graphs can be oriented horizontally or vertically • Put categories on one axis and numerical values on the other axis • Bar graphs are better than line graphs for comparing larger changes or differences in data among groups	
Line graph/ Scatter plot	$\geq$ 2 numerical variables	• Often used to show the value of a quantity over time • The independent variable is placed on the horizontal axis • The dependent variable is placed on the vertical axis • Make a scatter plot when the dependent variable is discrete	
Histogram	One continuous variable	• Allows for inspection of the data for its underlying distribution • Can spot outliers • Unlike a bar chart, there are no "gaps" between the bars (although some bars might be "absent," reflecting zero frequency) • Consider looking in section 2.3.2 for more on histograms	
Box plot (a.k.a. box whiskers plot)	One numerical variable	• Shows measures of spread and center • The box in the middle shows the "interquartile" range (i.e., where the middle 50% of the data lie) • A line segment in the box middle of the box indicates the median of the data • "Whiskers" indicate the max and min	

Table 3: Visualization types.

3.2 VISUALIZATION FOR EXTERNAL USE

Your modeling efforts matter most when you are able to explain them to others in a clear, succinct way. Visualizations can be a powerful way to communicate your data and results quickly and clearly.

The key concept to keep in mind is that when you create a visualization, it makes sense to YOU, but might not make sense to the person looking at it. In a report, you will not be there to explain the figure, so *you need to use words to explain what you want the viewer to see in the visualization.* Imagine you are introducing it to someone who hasn't seen it before. What should they look at first? How should they think about what they see? Should they compare two things? It might help to imagine describing the visualization to someone who can't see it.

The following is a quick checklist of items you should ALWAYS consider as you are preparing figures.

Visualization Checklist

☐ Is your graphic fully labeled?
  ○ Are the axes labeled?
  ○ Does the graphic have an informative title/caption?
  ○ Are units identified?
  ○ If needed, did you include a legend?

☐ Is the graphic readable?
  ○ Is the scale appropriate?
  ○ Are contrasting colors or textures used to represent different values in your output?
  ○ Is the font size large enough to be easily read?
  ○ Is the font style simple (like Arial or Calibri)?
  ○ Is the take-away message of the graphic clear?

☐ Does your graphic have extraneous or redundant information that can be eliminated?

☐ If you have several graphs side by side, do they present information consistently? For example, make sure the axes are consistent and that colors/ shapes mean the same thing as much as possible from graph to graph so that your audience can easily make comparisons, for example, associate blue triangles with category 2 hurricanes.

☐ Did you cite your source(s)?

Be aware that many people have some form of visual challenge, including color blindness. In order to make your visuals accessible, you need to think about how they would look if you printed them out in greyscale instead of color. You can use different shapes, textures and linestyles in addition to color so that the work is more useful for everyone. You also can make the font larger and more bold. A good rule in a paper is to make the font in your figures appear the same size as the font in the rest of the text. In a presentation, it matters how big your room will be and how large the projection will be. Keeping fonts large and clear is always safer. These principles will make your presentation clearer for everyone.

Saving graphics

Once you've polished up your figure, you'll need to copy it into your report or presentation. You may be able to "cut and paste," but you might want to think about whether it makes sense to save the figure or export the figure as a `.jpeg`, `.png`, `.eps`, or `.pdf`, for example. In MATLAB, you can save also a figure as a `.fig` file so that you can open it again with MATLAB and modify any features (for example, add units to the label on your axes) and then you have the option to export it as a variety of different file types to use when you want to insert it into a reporting document (write-up).

There are other situations in which you may not be able to save the graphic directly, even if you create it. In this case, consider using a screen capture tool (i.e., a screenshot). Many computers have built-in software (such as a "Snipping Tool") that allows you to highlight a portion of your screen that you want to share, and then save the image. On Macintosh computers, this same functionality is built in and can be accessed by simultaneously pressing command+shift+4.

Figures you find

It's also possible that you run across a figure that you want to share that you didn't create. You should still use the checklist above to make sure it's worth including. If you find EXACTLY the right graphic in a paper or article, you can include it in a report or presentation, as long as you cite it carefully to give credit where it's due.

Printed copy?

If you are creating a figure that will be included in a printed report, then, in addition to the checklist above, you might consider the following:

- Will your paper be printed in color or in black and white? Consider using dashed and dotted lines if the plot will be printed in black and white.
- Is your presentation accessible to a diverse audience? For example, a plot with multi-colored dots may be indistinguishable to a color-blind individual; consider using multiple shapes to identify different data sets.
- When you refer to your graphic, do you refer to "Figure 5" (good) versus "the figure below" (not so good)?
- Have you sufficiently explained the figure in your report? It's not enough to simply point your reader to a figure; you need a written explanation to accompany the figure and highlight the features that caused you to include it in the first place.

Presentation?

If your figure will be part of a presentation that will be projected, think about the following:

- If you'll be presenting on a monitor of some kind, then light text on a dark background works well.
- If your presentation will appear on a screen, you'll want to take into account how dark the room is. Light text on a dark background can work if the room is dark, but if the lights are on or if there's considerable ambient light, then dark text on a light background works better.

- Make sure your smallest font is big enough to be read from the back of the room.
- Some colors get washed out by a projector; it's really hard to spot a yellow line on a white background, for example.
- If at all possible, arrive early and go over your slides using the projector so you can work out any issues. Walk around to make sure figures are clear no matter where people sit in the room.

Remember the big picture

Create specific visualizations to support your ideas, and make sure your visualizations make your key ideas stand out. Then include supporting paragraph(s) that tell your audience what to infer from the visualizations.

And, finally, choose figures for inclusion in a report/presentation wisely. More isn't always better for your audience, who can get overwhelmed by too much information. Make sure you use only figures that illustrate and communicate important points. Within those figures make sure you include only the relevant information.

3.3 EXAMPLES: A PICTURE IS WORTH A THOUSAND WORDS, SOMETIMES

Here we give several examples highlighting and demonstrating the ideas from both internal and external visualization.

3.3.1 Example: Scale matters

We keep emphasizing that the way data is shown is critical, especially if you are going to draw conclusions from the graph itself. The *scale* used on your axes can play a role in what the graph looks like. Let's consider the following modeling scenario.

Suppose you take some medication and want to know when it will be out of your system. You look up information about the drug and estimate that at the end of every hour, half the previous amount remains. Let's let $C(t)$ be the percent of one dose of the drug that is in your body at time t in hours. Based on your assumption, if at $t = 0$ we have 100% of the drug, then after an hour we would have 50% and then after another hour we would have 25% of the drug and so on. This can be represented mathematically as $C(t) = 100 * (1/2)^t$. If we just plotted our function $C(t)$ over 24 hours, then we would get the top graph shown in Figure 30. It would appear that the drug is out of our system by about 6 hours (shown by the circle on the graph). Can we believe the graph? As time increases, intuitively we know the concentration is going to zero—so the concentration values will be very small in value as t grows. If we show the same values on a *log y* scale, we can see the amount of drug present based on powers of 10. This is shown in the lower graph of Figure 30. Here you can see that after 6 hours the amount of drug left is not as close to zero as it may seem in the first graph, although it is fairly small (about 1.5% is left according to the model itself). If you decided to only show the top graph, a reader might misinterpret the results of your model.

3.3.2 Example: Pie charts

Suppose we have data consisting of a puppy's weight over the first six months of life. Take a look at Figure 31, where we see how this data might be represented in a pie chart. When you look at this pie chart, what questions do you have about it? What is the meaning of the

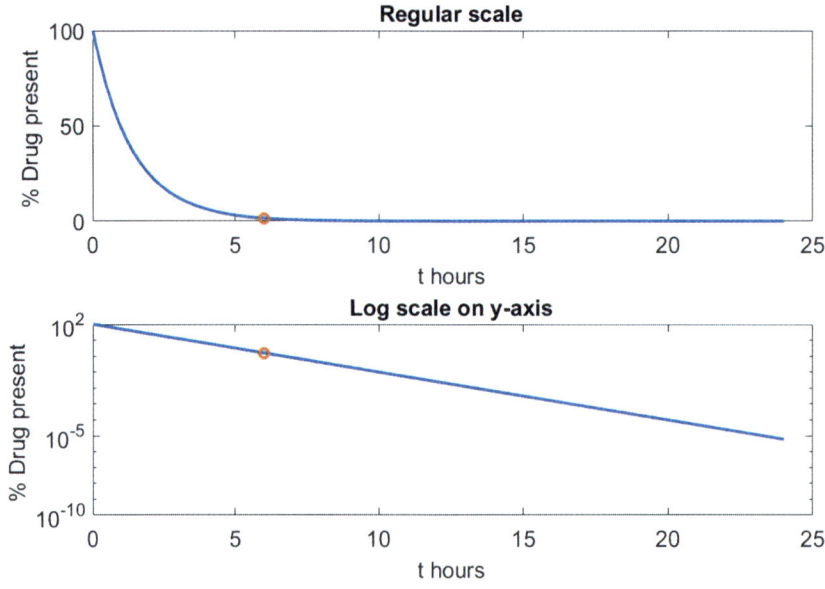

Figure 30: This shows the difference between using a linear or log scale on the y-axis. In the top graph, it is not clear how small the values are after say, 6 hours. They all look like zero. The bottom plot shows the actual magnitude of those values as powers of 10.

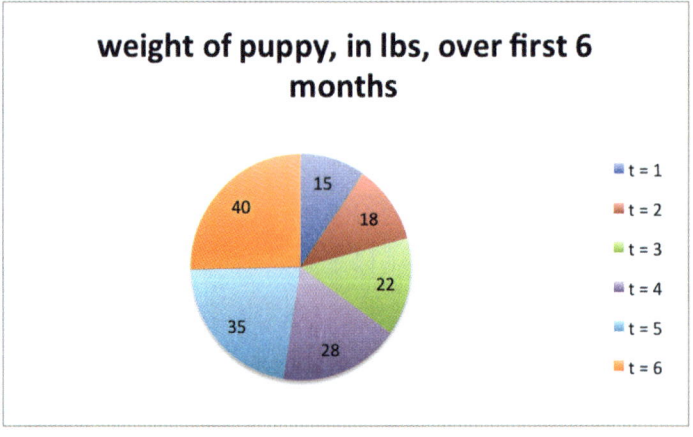

Figure 31: Pie chart example. It's not appropriate to use a pie chart to represent this data.

numbers floating in the wedges? What is the meaning of the green wedge with the number 22? What big, overall message do you get from this visualization?

Unfortunately, this pie chart is not helpful; it takes a long time to digest it and figure out what information it's conveying. Once you do finally figure out what the chart is showing, it still doesn't help interpret the data. For example, we might want to learn whether the puppy's growth rate is constant throughout the first six months, but this representation of the data makes it difficult to answer that question. In fact, *it would be easier to look at a table containing the data than to look at this pie chart!* The biggest problem with this particular representation is that pie charts are most useful when trying to show the relative sizes of parts of a whole.

Take a moment now to examine the pie charts in Figure 32. These are more appropriate for a pie chart than the puppy weight example was, because they help you understand about proportions of a renovation budget; they show parts of a whole.

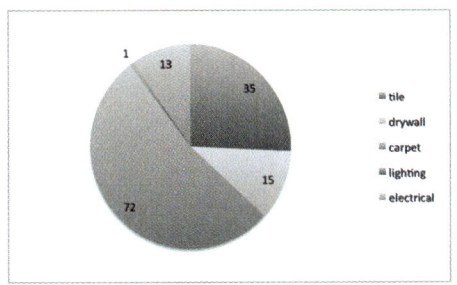 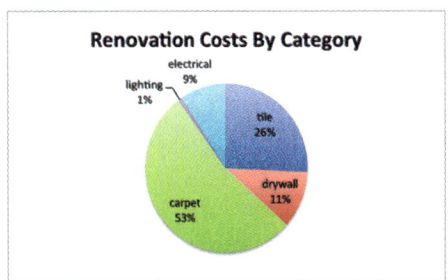

(a) This pie chart is a less-than-optimal way to show costs in a home renovation project.

(b) This pie chart is a much better way to convey the proportions of a home renovation budget.

Figure 32: Two pie charts that convey the exact same information about renovation costs.

While both of these show the same information, the chart in Figure 32b is better than that in Figure 32a at conveying the information for multiple reasons:
  • In Figure 32a:
    – The numbers don't add to 100, which is confusing.
    – There's no title.
    – There's not much contrast in the colors of the pie slices, so it might be hard to determine which label goes with which wedge.
  • In Figure 32b:
    – Percentages shown, so the numbers add to 100, as we expect when we see a pie chart.
    – There's a descriptive title.
    – Contrasting colors in pie slices help us easily distinguish between categories.

The pie chart in Figure 32a is a good starting point, and is exactly the sort of chart you might have for *internal use.* You can see the relative sizes of the categories and this might help you make modeling decisions. However, if you wanted to include this in a final report/presentation, you'd want to polish it up so it would look more like the pie chart in Figure 32b.

From these examples we see that you need to use the right visualization for the data type you have. Further, once you have chosen an appropriate visualization, you still might have to do some work to make it as useful as possible for your audience.

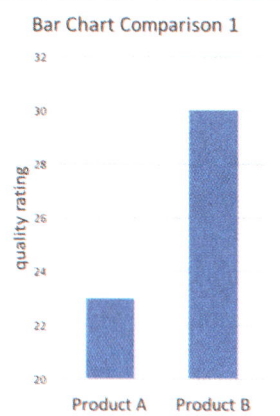
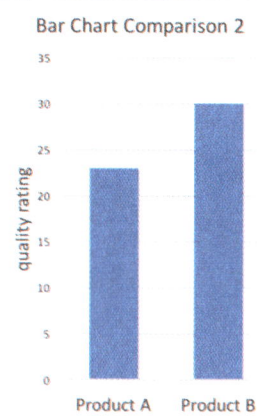

Figure 33: Bar chart example. These bar charts both represent the exact same data, but Bar Chart 1 is misleading because it exaggerates the difference between the two products.

3.3.3 Examples: Misleading with a visualization

One thing you need to be careful about is presenting data in a misleading way. Consider Figure 33.

Bar Chart Comparison 1 would lead you to believe that the quality rating of Product A is significantly lower than Product B. If you look closely, though, you'll notice that the bottom of the bars only go down to 20. Bar Chart Comparison 2 shows a more accurate depiction of the same data.

Bar charts aren't the only visualization that can be presented in a misleading fashion. Consider the plots in Figure 34. Which appears to attain a greater maximum value? Minimum value? Which one oscillates at a higher frequency? Carefully examine the plots before you answer!

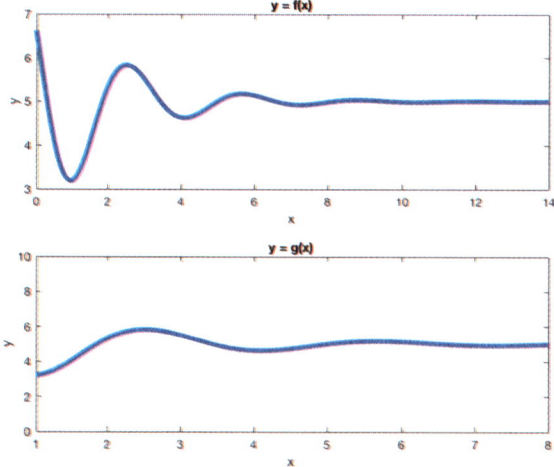

Figure 34: Line graphs example. The functions $f(x)$ and $g(x)$ depicted in the plots above are identical. However, the viewing windows are not the same, so they do not appear the same.

Does it surprise you to learn that the functions $f(x)$ and $g(x)$ are actually *the same* function? When two graphs are plotted next to each other (either side by side, or one directly beneath the other), you naturally want to make comparisons, and you may not notice that the axes are not the same. If you intend for your audience to make comparisons, make sure that the axes are scaled the same so as not to mislead.

If your audience determines that your visualization is misleading them, they won't trust your model or explanation. Keep this in mind when you are looking at others' work, as well; be careful when drawing conclusions from graphs/charts you find.

3.3.4 Example: Different visualizations highlight different aspects of data

We have mentioned that there may be multiple visualizations that are appropriate for a given set of data. Here we highlight the idea that you can tell different stories with different visualizations of the same data.

Consider the data in Table 4, which shows budgets and expenditures for three different cost centers of an organization.

Center	Q1 2016		Q2 2016		Q3 2016		Q4 2016		Q1 2017	
	Budget	Actual	Budget	Actual	Budget	Actual	Budget	Actual	Budget	Actual
1	1347	1200	2132	1980	2040	1976	1770	1618	3090	2938
2	2738	2300	1633	1245	4094	3845	1787	1523	3300	3152
3	2667	2210	3471	2980	2439	2398	3661	3566	1579	1264

Table 4: Cost center budgets and expenditures, in thousands of dollars.

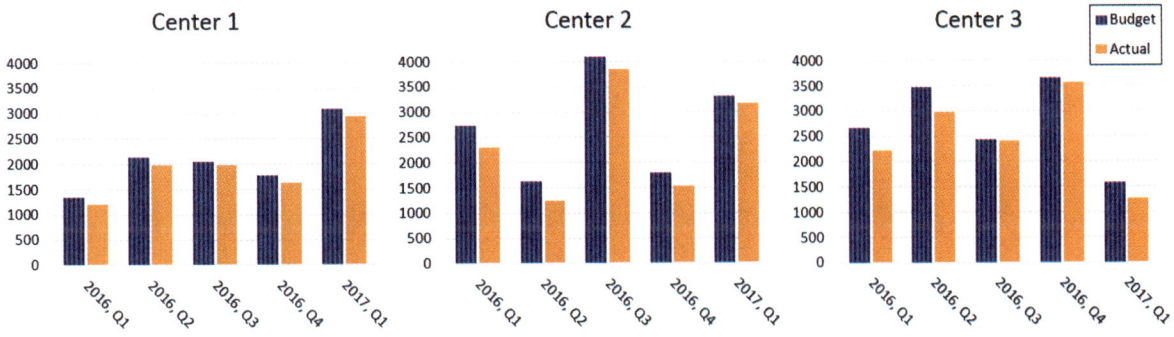

Figure 35: Cost center budgets and expenditures, in thousands of dollars.

Initially we might consider using bar charts to show dollars budgeted versus actual amount spent at each cost center during each quarter. This approach is shown in Figure 35.

While there is nothing wrong with this visualization, you may want to "see" various aspects of the data that could be represented better by looking at it a different way.

Focus: Comparing across time periods

If you want to study how accurately money is budgeted for spending over a set amount of time, you'll want to compare percentage of budget used across quarters. In this case, you might consider a visualization like that in Figure 36.

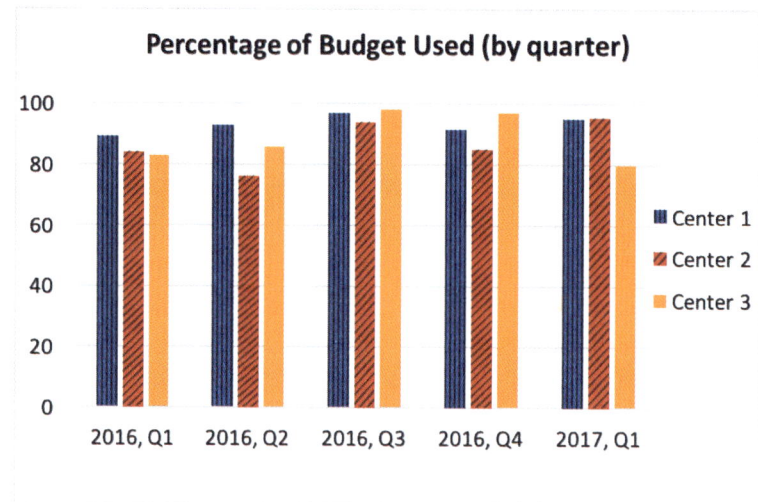

Figure 36: Percentage of cost center budget used by quarter.

Focus: Comparing centers

If instead you want to compare budgeting accuracy of individual cost centers, you might consider a visualization like Figure 37.

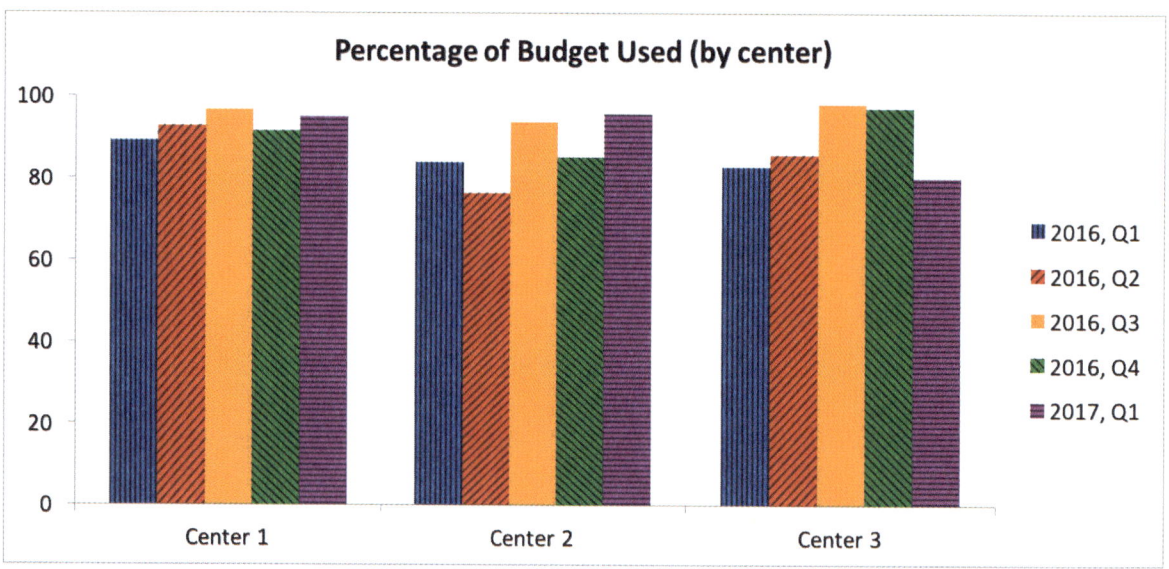

Figure 37: Percentage of quarterly budget used by centers.

There is no one "right" way to visualize your data; you should always consider which aspects of the information you want to highlight or focus on when creating graphics to share with an audience.

Note: Different colors are not enough.

You might have noticed that in Figures 35, 36, and 37 varying colors *and textures* are used to differentiate plotted values. Be aware that your audience may be color-blind or may encounter your graphic in a noncolor format. Using accessible visualizations increases the likelihood that your modeling outcomes will be understood by your audience.

3.3.5 Example: In (and out of) the pits

Following a speeding penalty that cost a NASCAR driver victory in a 2017 race [3], a team of student modelers investigated the effect of implementing different speed limits in Pit Road during NASCAR races. After the team did a careful analysis, they made a presentation to their peers that included Figure 38 as an example of the time decrease (or increase) top drivers may experience if driving at different speeds in Pit Road at Daytona International Speedway. Also included in the analysis is the effect of having a fast (or slow) pit crew.

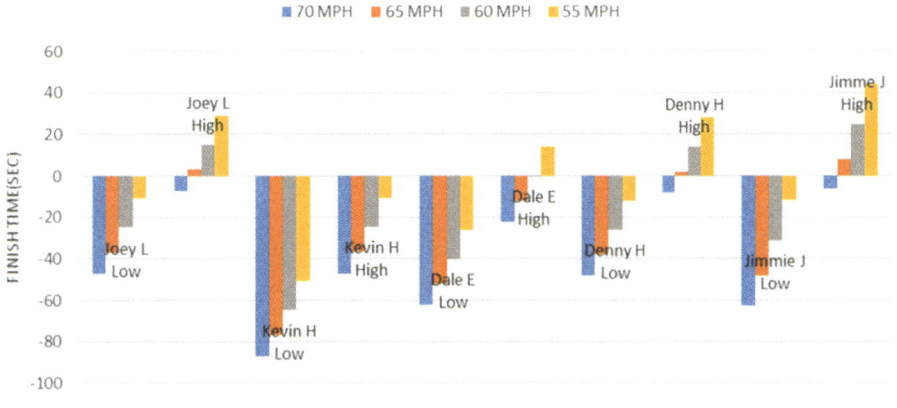

Figure 38: Even though this presentation slide satisfies many checklist items, it would be difficult for an audience to understand without significant explanation.

While Figure 38 satisfies many of the items found in the Visualization Checklist (p.43) the audience found it difficult to understand. Below are a few of the questions students asked—all about this graphic—following the presentation:

- Who are Joey L and Kevin H?
- What do High and Low mean?
- Who won? And who would (or could) win?
- What does Finish Time (sec) mean? (and why are some finish times negative?)
- What is the significance of 70, 65, 60, and 55 MPH?

Look at the figure and see if you can answer any of those questions.

The student modeling team was able to answer the questions, but *it took more time to answer these questions than had been allotted for their entire presentation*. This particular visualization did not help the audience understand the results of their model; rather, it created confusion.

The team would have benefited from addressing a few internal questions about their visualization prior to sharing it. That is, it's necessary to consider the checklist items noted earlier, but satisfying that list does not guarantee you've developed a great visual for communicating results. Consider this short list of <mark>questions to ask yourself before sharing a visual:</mark>

- Recall the question you are working to answer with your model (or submodel). Does your graphic provide an answer to that question?
- Identify (up to) three items that you want to highlight in your visual. Can an individual outside your modeling team identify these without your help?
- Can you explain the result (or partial result, as appropriate) in 15 seconds using your graphic as an aid? Or, if writing, can you explain the result using three sentences?

Let's revisit Figure 38 and answer the first question: Does the graphic provide an answer to the question being solved? This student modeling team was focused on answering the question of Pit Road speed in a general sense; as a result, who won was not nearly as important as understanding how final results could change. That is, including individual names distracted the audience so much that they were less concerned with the answer to the question. As a result, it would be reasonable to remove the names from the graphic and instead share them in narrative text, or use them as an example situation during a presentation. Additionally, the team incorporated two parameters into the same result; time spent driving on Pit Road, and the time it takes for mechanics to change tires and refuel the vehicle. That these two parameters were summed in Figure 38 wasn't clear to the audience and led some listeners to doubt their overall results. Instead, this team might consider spending time to highlight each parameter individually; in addition to providing an opportunity for more detailed analysis (e.g., is driving performance correlated with pit crew performance?), they will win the trust of their audience by being more transparent. Finally, many audience members didn't know if a negative time was good or bad. In this case additional language may help fix the problem. Explicitly define a value as "time saved…" (with units, of course) in order to clearly share these results. The table in Figure 39, while not as colorful as Figure 38, addresses many of the concerns the audience raised following this talk. In particular, it demonstrates that it may be possible for a driver to move up four spots if they are better at regulating (i.e., "maxing out") speed when entering and exiting from Pit Road.

NASCAR Speed Limit on Pit Road at Daytona International Speedway			
Speed limit (mph)	Pit Road driving time (sec)	Time saved if driving speed limit + 5 mph (sec)	Number of drivers that might have won using + 5 mph strategy*
55#	19.83	-	-
60	18.18	1.65	5
65	16.78	1.40	4
70	15.58	1.20	4
75	14.54	1.04	3

* 2015 Daytona 500
\# Current Daytona 500 pit road speed limit

Figure 39: A table is capable of sharing a lot of information in a clear and concise way. In addition to being more readable than Figure 38, this table clearly communicates the significance of a change in speed on Pit Road with respect to the overall result of the race.

Presentation components

Whether you are presenting your model as a quick summary at a career fair; a poster, presentation, or paper; or to a genuine client, most model communication includes these components (with visualizations throughout as helpful):

- Compelling descriptive title.
- Short motivating introduction that includes what question you answered, the context, who cares about the solution, and why the work is important.
- Background information.
- The model, including a justification of your assumptions, simplification, choices, and computations as well as a clear citing of your sources.
- Main conclusions.
- Analysis of when your conclusion is valid and what you might do with more time.

Be aware that, as the modelers, we can't know whether we have a clear, compelling presentation until we try it out on audience members unfamiliar with our work. We all tend to think we are being clear. A fresh perspective can help us improve our communication.

3.4 TAKE-AWAY MESSAGES

- Visualization of data is an important step in analyzing it and should not be overlooked. You may wish to make a plot, graph, or chart in order to:
    - summarize data,
    - make comparisons,
    - reveal patterns (i.e., correlations, relationships),
    - add interest to a report or presentation,
    - demonstrate points, and/or
    - facilitate understanding of information.
- Before you can determine the best way to visualize your data, you'll need to classify, and possibly sort, the data. Recall that data may be either categorical or quantitative, or a mixture of both.
- Visualizations can help you understand data you have found and/or can help you interpret the output of your model. Visualizations you produce for internal use need not be fancy; they simply need to help you make decisions (so they need to be useful and correct).
- When creating visualizations to convey your ideas to your others, it's important to think about whether the visualization will make sense to someone who is less familiar with your modeling question.
- In a report, you will not be there to explain the figure, so you need to use words to make clear what you want the viewer to see in the visualization.
- Refer to the Visualization Checklist as you are creating figures to be included in reports and/or presentations.

Concluding thoughts on visualization

It is easy to think that any visual is better than nothing. After all, the color, shapes, and textures make a presentation look nicer, and the presence of a visual can make a report look "complete" and professional. Graphs allow the readers to "see" and think about aspects of the model you have decided are important. We hope this chapter encourages you to use visualization as a powerful way to communicate quantitative and categorical information and ultimately make your model more understandable to others. At the same time, we hope you are careful in your choices so that your visualizations do not end up confusing or misleading your audience.

COMPUTATIONS

In the process of building your model, you will eventually get to the point where you need to find some numerical results. This chapter focuses on *computations,* and in particular on the decisions you'll be making: Do you need a computer, or can you get by with a calculator? If you need software, which package might be appropriate? When is the appropriate time to consider leveraging something more sophisticated, like a programming software? We will use two different examples (disease spread and farming choices) to show how the choices you make may lead to very different models and software needs.

4.1 EXAMPLE:
OUTBREAK?
EPIDEMIC?
PANDEMIC?
PANIC?
We all dread getting sick. Years ago, illness didn't spread very quickly because travel was difficult and expensive. Now thousands of people travel via trains and planes across the globe for work and vacation every day. Illnesses that were once confined to small regions of the world can now spread quickly as a result of one infected individual who travels internationally. The National Institutes of Health and the Center for Disease Control are interested in knowing how significant the outbreak of illness will be in the coming year in the U.S.

Note: This example also appears in the *M2GS2* handbook [4], so you can look there if you'd like to get more information on the decisions that led to the models that appear here.

Approach 1: You have a function to evaluate!
If you assume that 100 people initially have a disease and that the disease spreads such that the number of people infected doubles every week, then your model equation might be
$$I(n) = 100 \cdot 2^n,$$
where I is the number of people infected and n is the number of weeks that have passed.

If you are early in the modeling process, you might want to get just one value. For example, if you want to estimate the number of infected people after 12 weeks, you might just pick up a calculator and find that I(12) = 409600 people. If this number seems unrealistic, you may adjust your assumptions and adjust your model accordingly (for example, by changing the function or introducing a parameter).

If, instead, you are interested in knowing not only how many people are infected after 12 weeks, but also what the progression of the disease spread looks like throughout that time, you *could* use a calculator to find I(1), I(2), . . . , I(12). This might be a little bit tedious, but if you feel uncomfortable with technology, it will certainly get you an answer. If, however, you want to see how the disease spreads over 500 weeks, then you might be more inclined to use software, such as a spreadsheet program. You should definitely engage with technology when you would end up doing the same computations repeatedly.

You might start by setting up a spreadsheet as in Figure 40.

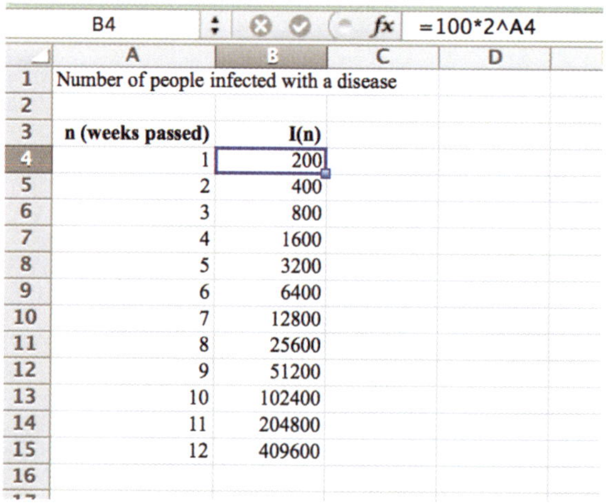

B4		fx =100*2^A4

	A	B	C	D
1	Number of people infected with a disease			
2				
3	n (weeks passed)	I(n)		
4	1	200		
5	2	400		
6	3	800		
7	4	1600		
8	5	3200		
9	6	6400		
10	7	12800		
11	8	25600		
12	9	51200		
13	10	102400		
14	11	204800		
15	12	409600		
16				

Figure 40: Initial layout of a spreadsheet for the disease model. Notice that cell B4 is selected in this screenshot, and we can see the formula for that cell in the formula bar at the top right of the figure.

Notice that the formula in the highlighted cell; that formula was "dragged" down to quickly generate the other output values (and it's equally easy to generate 10 values or 500). If you want more guidance on how to do this, you might do a quick internet search.

While you might consider a spreadsheet implementation for the sole reason that the actual computation went quickly once the spreadsheet was set up, one of the most powerful aspects of computing using a spreadsheet for computation is that it provides an easy way to vary parameters and see changes in the output. You can achieve this by putting the parameter values in their own cells and then referring to the cells, as in Figure 41.

C10 fx =$B$4*2^A10

	A	B	C	D
1	Number of people infected with a disease			
2				
3	doubling period (wks):	1.5		
4	initial number infected:	60		
5				
6				
7	n, number of	number of		
8	doubling periods	weeks passed	I(n)	
9	1	1.5	120	
10	2	3	240	
11	3	4.5	480	
12	4	6	960	
13	5	7.5	1920	
14	6	9	3840	
15	7	10.5	7680	
16	8	12	15360	
17	9	13.5	30720	
18	10	15	61440	
19	11	16.5	122880	
20	12	18	245760	
21				

Figure 41: Updated layout of spreadsheet for disease model, to include the ability to adjust parameter values.

Dollar signs (as shown in the formula for the highlighted cell in Figure 41 indicate that as the formula is "dragged" down, it should always reference the indicated parameter value. Once such a formula is created, you can simply adjust the parameter values (i.e., the doubling period and the initial number infected), and the entire spreadsheet will be updated. In fact, you can plot the output of the model (look at the Visualization chapter for more details), and even the plot will update immediately when you change a parameter value.

In general, the idea of seeing how the output changes when you vary parameter values is called *sensitivity analysis,* and it can be an important part of analyzing a model (see [4]).

Approach 2: Revised model, solution using symbolic computation

One of the downsides of the model used in Approach 1 was that it assumes that the number of people becoming infected continues to double over time, no matter how far into the future we look. In reality, we would expect disease spread eventually to slow down.

One way to account for disease spread slowing down is to think about how that rate changes. It might help to think about how the flu spreads through a school. Early on, when only one or two people have it, it spreads slowly. Later, when more people have the flu, then there are more people to spread it, so it spreads quickly. Eventually, however, almost everyone has it (or had it), so there aren't many people to catch it, and the rate of spread of the flu slows down again.

We can surmise from this behavior that the rate of spread of a disease depends not only on the number of people who have the disease (I), but also on the number of people in the population who don't yet have the disease (which is given by $P - I$). As we observe above, the rate at which the number of people are infected is also changing. We can use the symbol dI/dt to indicate the rate at which the disease spreads. Alternately, you may see the notation I' or $I'(t)$ to indicate the rate of disease spread.

If we use these assumptions, and assume a population size of 1000, then we might find a model of the form

$$\frac{dI}{dt} = k \cdot I(t) \left(10000 - I(t)\right).$$

For now, we'll just assume that the proportionality constant is $k = 0.3$, leading us to the following:

$$\frac{dI}{dt} = 0.3\, I(t) \left(10000 - I(t)\right). \tag{1}$$

(Again, to see the details of where this model comes from, see [4].)

If this equation looks unfamiliar, don't be intimidated! You really just need to know that the symbol dI/dt means the rate of change of $I(t)$, the number of people infected at any time t. In other words, equation (1), which is a differential equation, tells us how fast the disease spreads, but, unlike the model we had in Approach 1, it does not tell us *how many people* have the disease at any given time.

This particular differential equation happens to be solvable by "paper and pencil" techniques, but unless you've taken a course in differential equations, you may not know how to find the solution that way. When you have an equation or set of equations that you don't know how to solve, you might be able to leverage a symbolic computation tool (such as Mathematica or Maple or Wolfram Alpha), as shown in Figure 42.

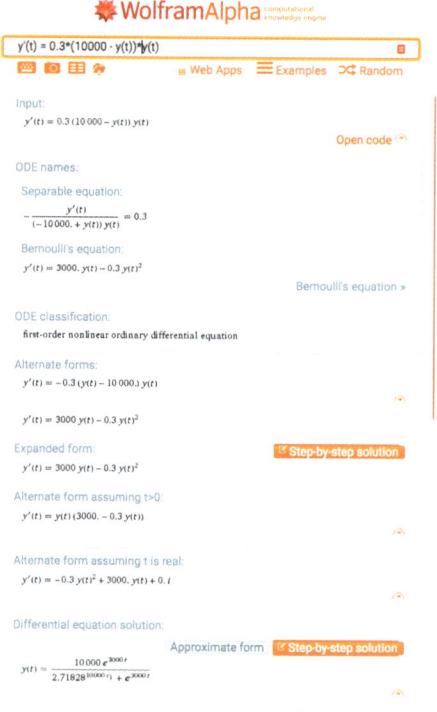

Figure 42: You can use a symbolic computation tool, such as Wolfram Alpha, to find analytic solutions for differential equations.

A couple of quick notes here about the solution using a symbolic computation tool:

- You'll note that we switched notation in order to get Wolfram Alpha to solve the differential equation (we switched from the variable I to the variable y, inserted function notation so it was clear that the independent variable was t, and changed the derivative notation from dy/dt to $y'(t)$). Computer algebra systems can be a little fussy about what notation they need... and unfortunately, each system is different and has its own set of "rules" or "syntax." It is sometimes helpful to do an internet search for specific examples showing how the software might be used.

- The solution Wolfram Alpha produced for this example,

$$y(t) = \frac{10000e^{3000t}}{2.71828^{10000c_i} + e^{3000t'}} \quad ,$$

is fairly simple, but sometimes the analytic solution produced by a symbolic solver is really complicated and spans several printed pages, which is not especially useful or insightful, in which case you might want to consider another computation option (see Approach 3).

Approach 3: Numerical solution

Let's continue with the differential equation,

$$\frac{dI}{dt} = 0.3\,I(t)\left(10000 - I(t)\right), \tag{2}$$

which is the same one we looked at in Approach 2. Let's suppose that this time we know in advance that we will want to know the answer at multiple values of time. If you have access to MATLAB or some other programming language, you can actually find a numerical solution. In terms of the example at hand, this would mean we could find the number of infected people over time without ever finding a formula for the function $I(t)$.

You can see an example of MATLAB code (using the function `ode45`) and the resulting plot of the numerical solution in Figure 43.

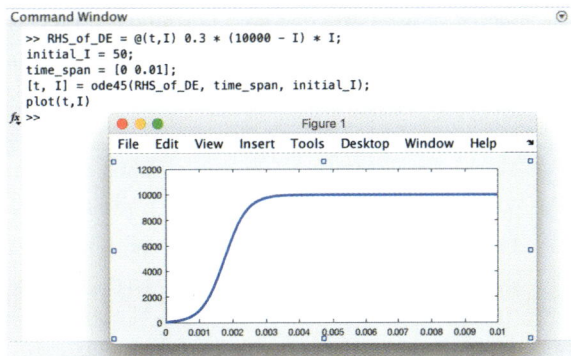

Figure 43: MATLAB and other software can be used to provide a numerical solution when finding a formula for a function is not necessary or practical. *Note: The plot here would definitely need to be improved if you wanted to include it in a report; head to the Visualization chapter if you need guidance!*

A couple of quick notes here about the solution using a numerical tool:

- You probably have no idea what MATLAB did to produce a solution! That's OK, for the most part, but it means you need to be really careful to check that your answer makes sense. In this case, for example, if MATLAB suggested that there would be a negative number of people infected, you should be suspicious. One of three things could be going on:

  - You may have a great model on paper, but you made a mistake entering it in MATLAB.
  - MATLAB correctly solved the model, but your model was flawed. You can adjust your model accordingly and then use MATLAB to solve.
  - Your model is reasonable, and you did implement it correctly in MATLAB, but you *still* are getting unreasonable results. This falls under the next bullet point.

- Each tool/command in MATLAB/Python/etc. has limitations on what kinds of problems it is good at solving. As in carpentry, you certainly need to own a hammer and it's an important tool, but a hammer is not helpful when the task requires a screwdriver. In MATLAB, `ode45`, for example, will only provide a reasonable solution for certain classes of differential equations. When you want to use a sophisticated tool (like a differential equation solver or an optimizer), you don't necessarily need to know the exact details of how the tool does what it does, but you do need to know whether the tool can even be expected to solve the kind of model you'd like it to. Read up a little bit about the function you plan to use to make sure it's appropriate for the problem you want to solve.

4.2 EXAMPLE: FARMING DECISIONS

Water you going to plant? Lettuce help you by using math!

Efficient water use is becoming increasingly important as periods of sustained drought and overuse of aquifers put available resources in jeopardy. In regions of intense agriculture, this problem requires immediate attention to ensure food security and a stable economy. However, farmers need to balance environmental concerns and changing water use policies with demand for certain crops and the ability to make a profit. For example, maybe some lucrative crops require significant irrigation, while other less profitable crops are in high demand. In addition, farmers may plant an entire farm and, depending on the growing cycle and effort required, be locked in to a particular crop portfolio for a few years (for example, raspberries have a two-year growing cycle). During that time frame, the price of water may increase drastically while rain events are scarce.

Farmers across the country have regional crops to choose from that will thrive in their area. They must select how much of each to plant. In California, a farmer may have to choose between strawberries, raspberries, and lettuce. In Nebraska, it may be potatoes, corn, and alfalfa. Suppose you are a farmer and have three possible crops to plant on your 100-acre farm. To generalize this complex system we will consider three crops, A, B, and C, each with different benefits and challenges. Suppose Crop A is high in demand and profitable but water intensive, Crop C has the lowest demand and lowest profit, but uses the least amount of water, and Crop B will be in the middle for all three properties. Create a mathematical model to determine what to plant for the next two years.

Developing an initial model.

There are lots of unknowns and many ways to think about this problem. What exactly is meant by high demand versus low demand? How long does it take for each crop to be harvested? How much water, exactly, does each crop require? How much money will be earned when the crop is harvested?

This is a little overwhelming! Remember that modeling problems are messy and not well defined, so it's totally normal to feel overwhelmed. In order to make some progress, it's reasonable to make assumptions so we can make a smaller, more tractable question. For our first approach to answering this question, let's start by not taking demand into consideration. Later approaches will incorporate demand, after we've developed results and intuition from the simpler model. You'll notice that we'll still introduce notation and values for demand, however, so that it can be used in later sections.

It looks like the following are unknown quantities:

- for each crop:
    - demand (i.e., the number of acres of your farm the open market would dictate should be dedicated to this crop),
    - number of acres you actually choose to plant with this crop (which might be different than demand),
    - number of months required before the crop can be harvested,
    - amount of water required for the crop (in acre-ft/acre),
    - sales price for the crop (dollars/box),
    - crop yield (number of boxes/acre);
- the price of water.

We'll also make some decisions about parameter values for our model, as shown in Table 5.

Crop	A	B	C
Yield (boxes/acre)	1000	1000	4000
Sales price ($/box)	4.00	3.00	2.00
Water use (acre-ft/acre/year)	3	2	1
Demand (acres)	50	35	15

Table 5: Model parameters for each crop.

We'll also assume that all crops can be harvested after four months.

With this set of assumptions, we still have several unknowns. We still don't know how many acres of each crop we will plant, so let's define

X_A = the number of acres of Crop A we will plant,
X_B = the number of acres of Crop B we will plant, and
X_C = the number of acres of Crop C we will plant.

If we want to write this in more compact notation, we might write

X_i = the number of acres of Crop i we will plant,

and it's understood that the i is an index that could be A, B, or C.

In all, we'll have the following unknowns:

D_i: the demand for Crop i (acres),
X_i: the number of acres of Crop i we will plant (acres), and
P_W: the price of water ($/acre-ft).

No farm can operate in the long term if it doesn't make money, so you'll need to determine how much money is generated when planting various crops. We'll think of total profit as the output of our model, and we ultimately want to maximize this.

Let's imagine you decide to plant 50 acres with Crop A. According to our assumed parameter values (Table 5), you would expect our income for this planting to be

$$(\text{50 acres of crop A}) \times \left(1000 \, \frac{\text{boxes}}{\text{acre}}\right) \times \left(\frac{\$4}{\text{box}}\right) = \$200000.$$

And since you would have to water the land, we would end up spending

$$(\text{50 acres of crop A}) \times \left(3 \, \frac{\text{acre-ft}}{\text{acre·year}}\right) \times \left(P_W \, \frac{\$}{\text{acre-ft}}\right) \times (\text{4 months}) = \$200 \cdot P_W.$$

Hence, our profit would be $\$200000 - \$200 \cdot P_W$.

Can you generalize this to get the profit if we plant X_A acres of crop A? Based on the sample calculation above, if we define Q_A to be the amount of profit generated when we harvest Crop A, then we have

$$Q_A = X_A \cdot 1000 \cdot 4 - X_A \cdot 3 \cdot P_W \cdot \frac{4}{12} \tag{3}$$

We want to do this for Crops B and C also, but it would be really nice if we could capture (3) in a compact form using subscripts, as we did before. In order to do that, we'll introduce the following parameters (for which we already have the values in Table 5):

Y_i: the yield for Crop i (boxes/acre),
W_i: the water usage for Crop i (acre-ft/acre),
S_i: the sales price for Crop i ($/box), and
D_i: the demand for Crop i (acres).

Now we are prepared to write (3) in a subscript form, as follows:

$$Q_i = \underbrace{X_i \cdot Y_i \cdot S_i}_{\text{revenue}} - \underbrace{X_i \cdot W_i \cdot P_w \cdot \frac{4}{12}}_{\text{cost of water}}.$$

(4)

Finally, we find our total profit by adding the profit for each crop:

$$Q = Q_A + Q_B + Q_C.$$

Note that you have only 100 acres available to plant, so you also have the following constraints:

$$X_A + X_B + X_C \leq 100,$$

We've now defined all of the variables and parameters and come up with an equation for total profit as well as two constraints. Let's gather up all that we've done for our initial modeling efforts, as shown in Figure 44.

Initial Farm Model

Variable	Definition	Units
X_i	The number of acres of Crop i we will plant	Acres
P_w	The price of water	$/acre-ft
Q_i	Profit generated from Crop i	Dollars
Q	Total Profit	Dollars

Parameter	Definition	Units
Y_i	The yield for Crop i	Boxes/acre
W_i	The water usage for Crop i	Acre-ft/acre
S_i	The sales price for Crop i	Dollars/box
D_i	The demand for Crop i	Acres

Assumed parameter values	Crop A	Crop B	Crop C
Yield (boxes/acre)	$Y_A = 1000$	$Y_B = 1000$	$Y_C = 4000$
Sales price ($/box)	$S_A = 4.00$	$S_B = 3.00$	$S_C = 2.00$
Water use (acre-ft/acre/year)	$W_A = 3$	$W_B = 2$	$W_C = 1$
Demand (acres)	$D_A = 50$	$D_B = 35$	$D_C = 15$

Model equations and constraints:

$$Q_i = X_i \cdot Y_i \cdot S_i - X_i \cdot W_i \cdot P_w \cdot \frac{4}{12},$$
$$Q = Q_A + Q_B + Q_C,$$
$$X_A + X_B + X_C \leq 100.$$

Figure 44: Farming Decisions Model: Definitions, parameter values, model equations, and constraints.

Approach 1: Exploration using initial model

If we reflect on our model (and recall again that we are ignoring demand for the moment), we can see that we actually have four unknown input variables: the number of acres of each crop to be planted (X_A, X_B, X_C) and the price of water (P_W). The output of our model is the total profit Q.

Now even more questions pop up! How do you look for a maximum value of a function when there are four input variables that you can vary? You might be more comfortable with techniques for finding maxima for functions of only one variable, like the classic $y = f(x)$, but it may not be clear what to do when you vary more than one, much less four things.

Explore the parameter space

While there are computational tools that can search for maxima (and we'll discuss some of those in the next couple of sections), there are some things you can do using almost any form of technology. Here are some options you might consider.

• Set the price of water to some constant value, and then choose combinations of low, medium, and high values for each of the X_i variables, as shown in Figure 45.

F12   fx =ROUND(A12*$B$5*$B$6 – A12*$B$7*$F$5/3,2)

	A	B	C	D	E	F	G	H
1	Initial Farming Model							
2								
3	Parameter values:							
4		Crop A	Crop B	Crop C		price of water:		
5	Yield	1000	1000	4000		50		
6	Sales Price	4	3	2				
7	Water Use	3	2	1				
8								
9	Input Variables:			Output:		Intermediate output values:		
10	X_A	X_B	X_C	Total Profit (Q)		Q_A	Q_B	Q_C
11	100	0	0	$395,000.00		395000	0	0
12	50	50	0	$345,833.33		197500	148333.33	0
13	0	100	0	$296,666.67		0	296666.67	0
14	50	0	50	$596,666.67		197500	0	399166.67
15	0	0	100	$798,333.33		0	0	798333.33
16	0	50	50	$547,500.00		0	148333.33	399166.67
17	50	25	25	$471,250.00		197500	74166.67	199583.33
18	25	50	25	$446,666.66		98750	148333.33	199583.33
19	25	25	50	$572,083.34		98750	74166.67	399166.67
20								

Figure 45: Exploring the parameter space for the initial farm model.

Based on these computations, you might conclude that to achieve the maximum profit you should plant all 100 acres with Crop C. Keep in mind that this process—choosing several sets of parameters—*does not guarantee* that you have actually found the parameters that achieve a maximum value for the output function, but you can find the maximum among your sets of parameters. Sometimes this process does help you identify *the* maximum, but even when it doesn't, it can still provide valuable information and might help you develop intuition or think of other parameter combinations to try.

• Set all the crop amounts to their demand values (i.e., set $X_i = D_i$) and vary only the price of water; this will generate a line, as shown in Figure 46.

Since all of our equations are linear in this model, you might be unsurprised by the linear plot in Figure 46. The idea of fixing all of the variables except one and then varying just that one

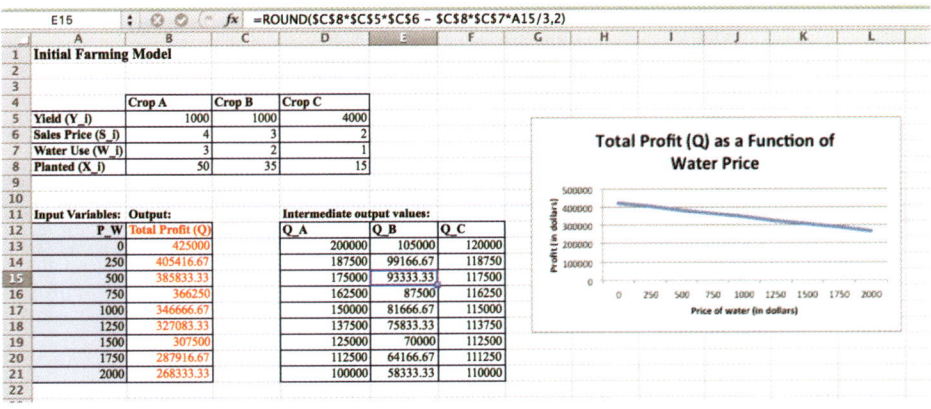

Figure 46: Profit as a function of only one variable.

remaining variable can actually be quite enlightening for nonlinear models.

- If you assume that you will plant every acre with some crop, then you know

$$X_A + X_B + X_C = 100,$$

so we can eliminate one of the variables, say X_C, by writing it in terms of X_A and X_B:

$$X_C = 100 - X_A - X_B.$$

Hence, we now have only three input variables. And if we make an assumption about the price of water (i.e., set P_W to a constant value), then we can vary the number of acres of the other two crops and see the effect on profit. As we mentioned before, all of the equations in this model are linear, so if we were to visualize this, we would see a plane.

For each of the above situations, software that can do many computations at once is clearly a benefit over evaluating the model for different values in a calculator and writing them all down. While these were implemented using spreadsheets, essentially any of the tools we are discussing in this book (spreadsheet tools, MATLAB, Mathematica, Python, etc.) would work.

Revisiting the model

During the initial round of modeling, we omitted some of the modeling considerations so that we could cut down on the (already considerable) number of variables and parameters. In particular, we never took into account the *market demand* for the crop. We assumed that if we planted it, we could sell it, which may not be a reasonable assumption. What would happen if, for example, ALL farmers in California figured out that it was most profitable to grow lettuce, so they completely stopped growing strawberries and raspberries and decided to grow lettuce exclusively?

In reality, farmers need to take into account what consumers want. We have a variable for demand:

$$D_i = \text{the demand for Crop } i \text{ (acres)}.$$

Assuming that farmers need to plant close to what is dictated by the market demand, we might also want to define a variable to indicate the difference between how much of a crop is actually planted (X_i) and how much demand there is for the crop (D_i). One way we might define the *deviation from demand* for each crop is

$$Dev_i = |X_i - D_i|,$$

and then the total deviation from demand is

$$Dev = Dev_A + Dev_B + Dev_C.$$

Approach 2: Optimization

Upon reading this problem, you may think you need to help the farmer decide the "best" set of crops to plant. That is called an optimization problem. For example, what should they plant over the two years to *maximize* profit while trying to not to stray too far from the demand? It would be difficult to use the above approach to do this because you would need to evaluate the models for many different combinations of crop pairs. Optimization is a way to pose a problem and then try to find a maximum or a minimum in a direct way, often using algorithms.

Posing an optimization problem requires identifying decision variables (in this case, the number of acres of each crop to plant: X_A, X_B, X_C) and developing a mathematical model that describes your goal (called an *objective function*). One goal here is to maximize the objective function given by the profit model above, but some other "rules" are needed to keep our model realistic and to account for the market demand. These rules are referred to as *constraints*. For example, you need to be sure to only use 100 acres, since that is the size of your farm. This is the constraint we saw earlier,

$$X_A + X_B + X_C \leq 100.$$

There are other considerations in this scenario that may be important to you such as minimizing water usage (although it contributes directly to the profit) or possibly providing jobs for the community. It would be up to the modeler to decide how to interpret the question and develop his or her own concise *problem statement*. We will consider three very different interpretations of this scenario, both with their own unique assumptions, and show how the appropriate software tools can provide choices for this farmer.

Optimization with Excel

Suppose we pose the problem as maximizing the total profit—how do we account for market demand?

One approach is to make the *assumption* that, for each crop, we do not want to change what we plant by more than 20% of what the consumer demand is. Mathematically we can pose this as $|Xi - Di| < (0.2)D_i$. So for Crops $i = A, B, C$ we have

$$|X_A - 50| < 10, \quad |X_B - 35| < 7, \quad |X_C - 15| < 3.$$

We can use algebra to expand the inequality and absolute values to get bounds on our decision variables, as follows:

$$40 \leq X_A \leq 60, \quad 28 \leq X_B \leq 42, \quad 12 \leq X_C \leq 18.$$

Now that you have an objective function and constraints, how do you solve the problem? Which software tool or built-in solver can you apply? Now is a good time to take a step back and examine your model and the constraints and try to *classify* your model. That classification can help you pick an appropriate software tool. Think about the following questions:

- Is the problem linear or nonlinear?
- Are the constraints inequalities or equalities?
- Are the constraints nonlinear or linear?
- Are the decision variables integers or real numbers?

For this model, you have a linear objective function (since our profit function depends only on your decision variables by multiplying by a constant and adding) and linear inequality constraints.

Excel has a Solver package that is an easy add-in that comes equipped with optimization tools that are suitable for this problem. If you didn't know about those tools in advance, but know you would like to use Excel because you are most comfortable with it, you may first look at the optimization tools available in Excel and then decide how to formulate your optimization problem. We'll discuss the mechanics of using this particular tool, but remember that you can always do an internet search to get help with unfamiliar tools.

To solve an optimization problem in Excel, you can organize your spreadsheet in any way you choose. You simply need to assign certain cells to hold the decision variables, and then create a formula in a cell to calculate your objective function value. You will also need to create a formula for constraint functions or bounds on your decision variables. When you call on the Solver tool, a pop-up appears, and you can fill in the values and then execute the solver, as shown in Figure 47.

As we see in Figure 47, we can maximize profit and stay within 20% of demand by planting 40 acres of Crop A, 35 acres of Crop B, and 18 acres of Crop C (the values in Column E).

With Excel, once your model is well-defined, the pop-up for the Solver makes getting a solution fast and simple. With this tool in hand, you can change various model parameters in the spreadsheet and run the solver again to see the impact on the solution, making this tool even more helpful.

If your optimization problem is nonlinear or has complicated constraints, there are certainly a wide range of other tools out there. We will discuss some of them below on this simple example.

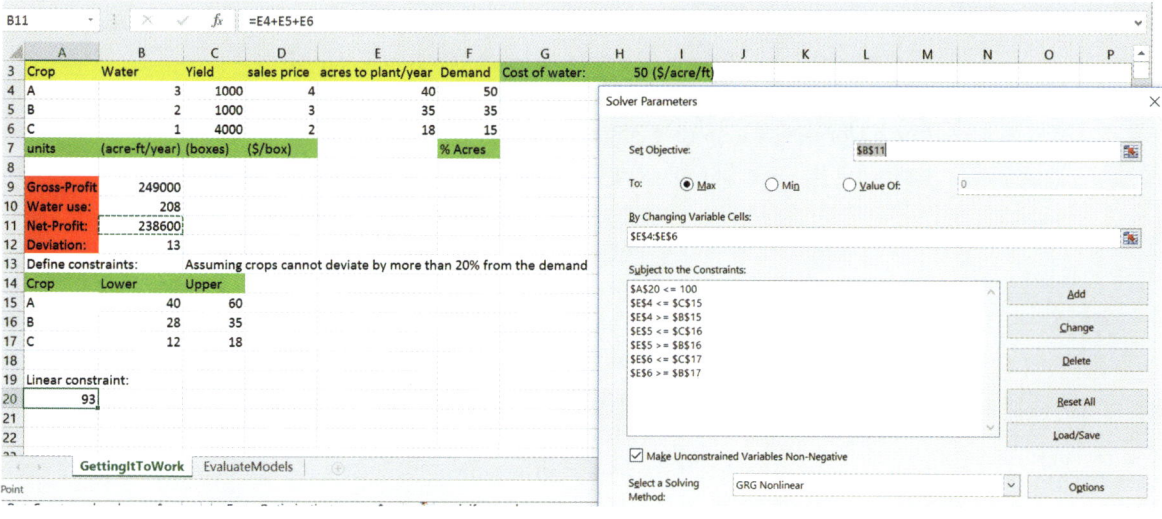

Figure 47: Possible organization of an Excel spreadsheet to use the optimization tool in the Solver add-in. The solver will fill in the values for how much of Crops A, B, and C to plant in Column E.

Optimization with a sustainability constraint

Because sustainable water practices play a critical role in many communities, water policies could be included in your model. This can be done by removing the water component from your profit function and placing a restriction on the amount of water used. This adds a different constraint to your optimization problem formulation of the form

$$W = W_A X_A + W_B X_B + W_C X_C \leq W_R, \tag{5}$$

where W is the total water used by crops A, B, and C, and W_R is the restriction you can impose on the water use. Our new profit function would just be the part of Q that accounts for the revenue,

$$\underbrace{R_i = X_i \cdot Y_i \cdot S_i}_{\text{revenue}}, \tag{6}$$

and we would get the total profit by adding the profit for each crop:

$$R = R_A + R_B + R_C.$$

So this model would seek to maximize the revenue R subject to the sustainable water use constraint and the acreage constraint (cannot plant more than 100 acres). MATLAB also has a wide range of built-in optimization tools. In fact there is an Optimization Toolbox that contains state-of-the-art algorithms that can tackle problems with hundreds of decision variables and constraints that can be highly nonlinear and complicated in nature. Since the profit model we are considering here is linear, and so are the constraints, linprog.m is one possible solver. Even if you didn't know about this program before, you do know that your problem is a linear optimization problem, so an internet search will point you in the right direction and even lead you to video tutorials about how to use this solver. Being able to classify your problem is key in identifying the right tools.

One way to learn how to use this built-in solver is by typing "help linprog" in the MATLAB workspace. Figure 48 shows the start of the explanation MATLAB provides.

One major observation that may throw you off at first is that we are hoping to *maximize* the profit, but this tool, like many optimization tools, is for minimization. We can get around that with a trick—by minimizing the negative of our function! (Think, for any function $f(x)$, the minimum of $-f(x)$ is the maximum of $f(x)$.)

If you look at the Help file in Figure 48, you can see what inputs and outputs the linprog.m function has. The function requires several inputs: you give it the *coefficients* of your decision variables from the objective function in a vector f, A and b for the linear inequality constraints $Ax \leq b$, Aeq and beq for the linear equality constraints $Ax = b$, and lower and upper bounds on your decision variables. If you have multiple linear inequality constraints, those would go into a matrix A and the right-hand side would be a vector b.

```
>> help linprog
 linprog Linear programming.
    X = linprog(f,A,b) attempts to solve the linear programming problem:

              min f'*x     subject to:    A*x <= b
                 x

    X = linprog(f,A,b,Aeq,beq) solves the problem above while additionally
    satisfying the equality constraints Aeq*x = beq. (Set A=[] and B=[] if
    no inequalities exist.)

    X = linprog(f,A,b,Aeq,beq,LB,UB) defines a set of lower and upper
    bounds on the design variables, X, so that the solution is in
    the range LB <= X <= UB. Use empty matrices for LB and UB
    if no bounds exist. Set LB(i) = -Inf if X(i) is unbounded below;
    set UB(i) = Inf if X(i) is unbounded above.

    X = linprog(f,A,b,Aeq,beq,LB,UB,X0) sets the starting point to X0. This
    option is only available with the active-set algorithm. The default
    interior point algorithm will ignore any non-empty starting point.
```

Figure 48: MATLAB's Help response when trying to understand how to use linprog.m.

Using our revenue model and the values for yield and sales price for each crop, we seek to maximize

$$R = (1000) * (4)X_A + (1000) * (3)X_B + (4000) * (2)X_C, \tag{7}$$

so the entries in f are –4000, –3000, –8000 (since we are finding the minimum of –R).

To enforce our sustainability constraint, let's assume we can use at most 235 acre-ft/acre of water per year. This gives

$$3X_A + 2X_B + X_C \le 235, \tag{8}$$

which incorporates the water use for each crop. These coefficients (3,2,1) are the entries in A and 235 is b. To account for our demand constraint, we will require that the amount of each crop be greater than zero but not exceed the upper bound derived in our previous version, and we set these values in the MATLAB variables lb and ub. Also, we will assume we want to use all of our farm, requiring the equality constraint $X_A + X_B + X_C = 100$. This means we set the entries in Aeq to 1,1,1 and beq to 100.

Once you have those model parameters properly defined in MATLAB, then you can call the solver. It is critical to understand how to do this, and most software tools, such as MATLAB, will have examples online to help walk you through the process. Below is the exact syntax entered into the workspace in MATLAB:

```
>>f = [-4000,-3000, -8000];
>>A = [3, 2, 1];
>>b = 235;
>>Aeq = [1, 1,1]
>>beq = 100
>>lb = [0,0,0];
>>ub = [60,42,18];


>>[X, FVAL] = linprog(f,A,b,Aeq,beq,lb,ub)
```

The solution you get when solving this formulation is $X_A = 53$, $X_B = 29$, and $X_C = 18$, which is different from our previous example, as expected since you are using a different underlying model.

Some thoughts on this approach:

• There may be an initial learning curve to overcome when you use a built-in solver in any software package, so it is likely that you will need to look up tutorials or read over documentation on how to use it correctly. Often, solving an easier problem that you already know the answer to assures you that you are using it correctly!

• For this problem, there may be some model parameters that you want to change to better understand the strengths and weaknesses of your approach and see how they impact the number of acres the farmer plants. To this end, it may be tedious to redefine the coefficients and model parameters (which you see from Figure 48 must be provided in an array). To keep from having to retype these in your programming workspace, it may help to embed this call to the solver in a larger program. We talk more about this in the chapter on Programs and Simulation.

• Although we showed how to solve this problem with MATLAB, there are many varieties of built-in optimization solvers in Python, Mathematica, and elsewhere.

Optimization using trade-off curves

For this model, and with many modeling questions, there are many competing goals. You'd like to make the most money, use the least amount of water, and meet market demand. It's impossible for all three of those to goals to be met simultaneously, but looking at different scenarios before making a commitment to planting would be valuable so that you can make informed decisions.

If you really dig into some of the optimization options in MATLAB you may stumble onto something called "multi-objective optimization," which is an attempt to consider several objectives, possibly competing with each other, to try to provide the user with a set of possibilities so they can make decisions based on their own values. For example, if you are working with an industrial partner, they may be more interested in meeting the market demand. If instead you are partnering with a policy maker, they may highly value sustainable water use.

This farming question could be formulated to try to maximize profit, minimize water use, or minimize deviation from the demand using the models proposed above. If you can classify this formulation as a multi-objective problem, you will once again find many tools available to you! The `gamultiobj.m` solver in MATLAB uses a multi-objective genetic algorithm to approach this problem. The details of the underlying algorithm are beyond the scope of this book, and the good news is, you really only need to know that you have chosen an appropriate solver and how to use it (and how to interpret the results). Again, the Help menu in MATLAB and/or an internet search can help you figure out how to use the function.

Figure 49 shows a trade-off curve comparing the water objective and the deviation from market demand. The set of points (stars) is called a "Pareto front" and is the final result of the optimization. Since we have competing goals and there is no one single plan that can both minimize water usage and meet market demand, the solver *returns a set of points* so that a user can choose a planting strategy that meets their own personal goals. Each star corresponds to design point (i.e., the amount of Crops A, B, and C to plant). A person can choose whether they want a strategy that is better for water conservation or one that stays close to the demand, or somewhere in between. For example, as you can see in Figure 49, notice that choosing a point with low deviation from demand (i.e., those points with *y*-coordinate close to zero—which happen to be on the bottom right in this figure) means you have to use a lot of water (because their *x*-coordinates are large).

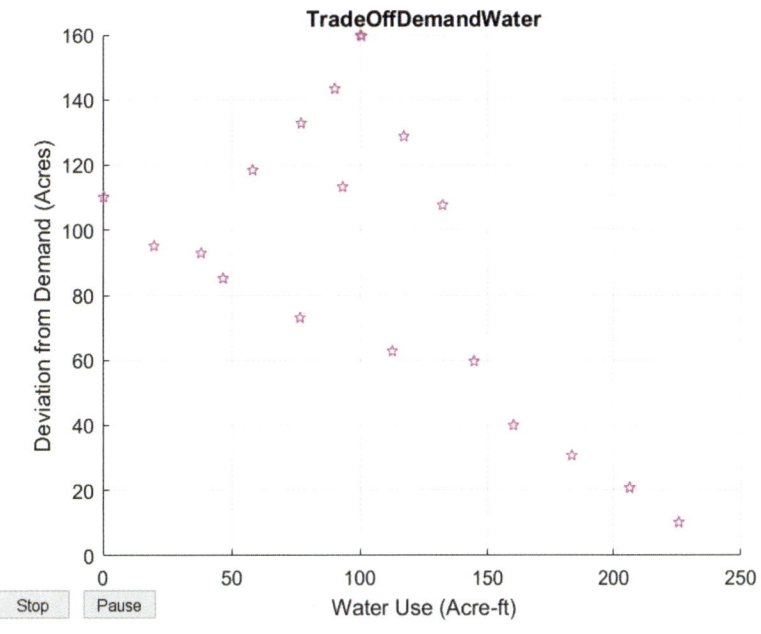

Figure 49: Trade-off curve generated using `gamultiobj.m` to help guide a farmer in what to plant. Each star represents a different design point (amounts of Crops A, B, and C).

Approach 3: Optimizing on a more sophisticated model

The previous approaches completely ignored a major part of the original real world question. Crops A, B, and C have specific growing seasons! For instance, we might consider a set of crops such that Crops A and B are in the ground for four months and Crop C is in the ground for 8 months. A farmer may want different amounts of Crops A and B in the ground after the first harvest, or even to let some of the land remain empty. None of the previous models allow for this flexibility. To account for this, you could write a computer program that steps in time and tracks what land is available while calculating the profit, water used, and how well the demand is being met, and then the farmer could make a decision. This is beyond the scope of this computations chapter and really moves into the realm of simulation and programming. Continue on to that chapter to see how it's tackled!

4.3 TAKE-AWAY MESSAGES

- You'll want to engage with technology to handle a larger number of computations.
- Classifying your model can help you look for an appropriate software tool. Also be cognizant of which values are your parameters and which are decision variables.
- If there are multiple decision variables and/or parameters, use technology to help you quickly explore results and gain intuition about your model.
- Using technology can also facilitate a sensitivity analysis, where you vary model parameters and see how much the output changes.
- If you have an equation or equations and you don't know how to solve, you might consider using a symbolic computation tool.
- You might consider a numerical solution when finding a formula/function is difficult and/or not enlightening.
- Optimization tools can be useful in answering many modeling questions. An internet search can help you find the right one for the software you are using.
- Whenever you use a computational tool, you need to be aware of the limitations of the tool and critically examine the solution it provides. A given computational tool won't work for every situation and unfortunately may not always give errors to alert you when you use it incorrectly.
- Using the software to solve an easier problem (that you know the answer to already) is a good way to be sure you are using the tool correctly!
- Reading over documentation and example problems and maybe even watching video tutorials is likely necessary to learn how to use a new tool. Be patient and pay attention to details to be sure you are using the tool correctly.

Concluding thoughts on computations

Even though the word "computations" seems straightforward, when it comes to modeling there may be more involved than just simple arithmetic. Having to do repeated calculations, solve a nonlinear equation or system of equations, or apply a mathematical idea to a large data set often cannot be done with a pencil and paper. The aim of this chapter was to provide guidance in choosing the right computational tool and to encourage you to explore options. When doing the computations becomes disorganized, overwhelming, or inefficient, creating a computer program to automate computations may help. That is the topic of the next chapter.

PROGRAMMING AND SIMULATION

We need to say up front that this chapter is a bit more advanced than the others, but want to emphasize that programming is not just for the elite. Anyone can become a good programmer with practice. We hope that you will have the confidence in yourself to give it a read and even take a first try at programming. In mathematics, when you are reading complicated material, it can help to scan through once just to get an idea of what is there and then go back and read again more slowly. You also might need to open up a computer and try things as you go, rather than just reading. It also probably won't be as easy as in some other chapters to jump around or start in the middle.

In the Computations chapter we discussed how software can be used to evaluate your model or find a solution. Throughout, we discussed that when computations may become tedious because you have to repeat them for different parameters, you may want to create a simulation. In this section we take computations one step further and introduce writing short programs or building simple simulation tools.

Not all computer programs are simulations, but simulations rely on building computer programs and doing "coding." The word simulation refers to the "imitation of a process or situation." In the context of this book, it means using computer programming to model a process, usually to study the behavior of a system. Simulation relies on having a model that represents the scenario you are interested in, then using software to generate output that can be interpreted as an imitation of that scenario. The scenario usually (but not necessarily) develops as time passes. Parameters can be changed to gain insight into the situation you are trying to represent and understand.

Simulations are often used when real experiments would be impossible, dangerous, or too costly to perform. Or you may want to use data that is unavailable to you, so a simulation can help generate that data. Other times you might develop some rules about the interactions between members of a population, or set some initial conditions and create a simulation to see how that population evolves over time (e.g., predator-prey relationships, or students and sports-related injuries).

As you begin to build/solve your model using software tools that allow for writing computer code, it usually becomes clear that a "larger" program may help you organize your subroutines, especially if you are calling a series of built in tools in a row. You usually can take various subroutines (or functions you may have already coded up) and then connect them in a bigger program to execute an entire sequence of commands such as: using a solver to get a solution, then evaluate the solution, perform a sensitivity analysis, and generate plots—all at once! For example, Figure 50 shows some questions that may come up that would lead you to put all those parts of the solution process into one program. We will follow this simulation decision flowchart in one of the examples below.

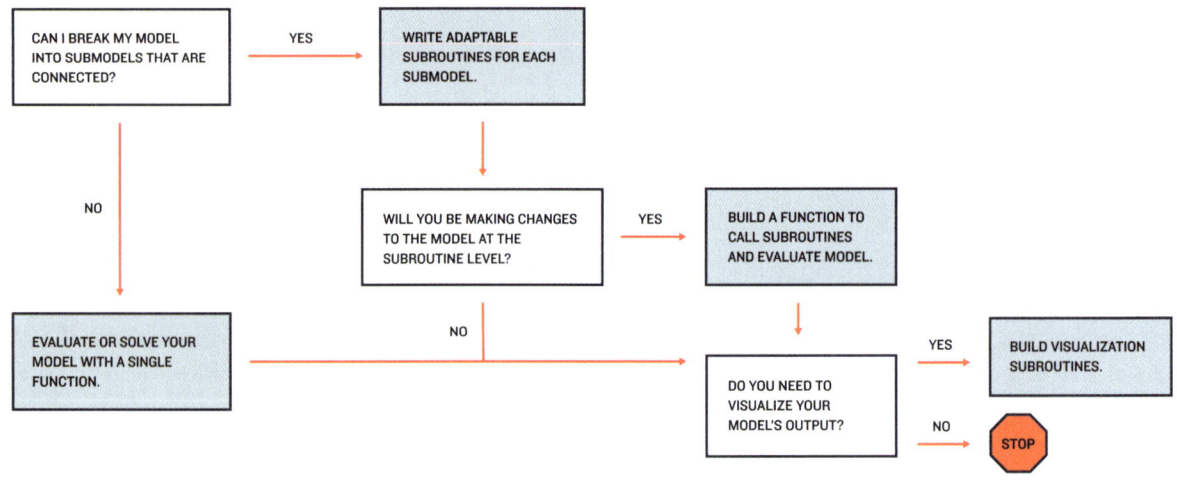

Figure 50: As you are using technology to build your model, you may get to a point where a larger program would help. For example, the questions may or may not lead you to creating a more detailed software tool.

In order to learn how to write computer code, you have to figure out how exactly the software interprets the set of instructions you write—this is no small task! The good news is that once you have figured out how to write instructions in one language, you can often use those skills to help you figure out how different programming software works.

5.1 EXAMPLE 1: SIMULATING STUDENT PROFILE DATA FOR THE LUNCH CRUNCH MODEL

Let's focus on the 2014 M3 Challenge Question, called Lunch Crunch [1]. In particular, the first part of the challenge asked:

Develop a mathematical model that takes as input a student's individual attributes, and outputs the number of calories that a student with those attributes should eat at lunch.

The second part was worded as follows:

> Now that you've identified attributes that affect caloric needs at lunch, create a model to determine the distribution of U.S. high school students among each of these categories. If every student eats the standard school lunch, what percentage of students will have their caloric needs met at lunch?

You might develop a sensible model or models to answer these questions, but eventually you'll want to "plug in" some students to see the output of your model. You would ideally have information about individual student attributes (including, possibly, height, weight, body mass index, how active the student is, etc.) for EVERY high school student in the U.S. Unfortunately, no such database exists. What you might do instead is use statistical tools to *simulate* school populations with students who have specified physical properties/caloric needs. You could then use those "representative" students as input to assess the validity of your model by seeing what fraction of the simulated student population has their caloric needs satisfied by the lunches your model recommends (the lunches would come from another model or simulation)

What describes a student's caloric needs?

What are the important characteristics of the students in the modeling problem that we could use generate a representative population? After reading the problem statement and discussing it as a team, you might decide that for each student you would like to have data on their:

- **Activity level** from [5]: sedentary (factor 1.2), mild (factor 1.375), moderate (factor 1.55), heavy (factor 1.7), or extreme (factor 1.9).
- **Grade:** 9, 10, 11, or 12,
- **Body mass index (BMI):** from 18 to 30.

Which distribution should I use?

In the Statistics chapter, recall that we mentioned a few of the ways data sets may be *distributed,* such as uniform, normal, or bimodal. When you simulate data using random values drawn from a distribution, you need to make assumptions and state your reasoning for your choices, even if your reasoning is that you are choosing the simplest option because you need a starting point.

For example, if you wanted to generate body mass index (BMI) ranging from 18 to 30, you might use a uniform distribution because you think any of those BMIs are about equally likely. On the other hand, if you are generating BMIs from 15 to 35 you might choose a normal distribution because you think the smallest and largest values are less likely. This means you are assuming that most people have a BMI of around 25 and fewer people have BMIs with larger and smaller values.

Here is a simple way to generate uniformly distributed numbers between 18 and 30. Let R be a random number between 0 and 1. When you multiply R by 12 (i.e., 30 − 18) and then add 18, the outcome is a number between 18 and 30. Try it! Pick a number R and then compute $R * (30 − 18) + 18$.

In Excel, the command `RAND()` generates a random number between 0 and 1 from a uniform distribution. To generate the five activity levels you can use nested IF statements:

```
=IF(RAND()<0.2,1.2,IF(RAND()<0.4,1.375,IF(RAND()<0.6,1.55,IF(RAND()<0.8,1.7,1.9)))).
```

The `RAND()` command is going to spit out a number between 0 and 1. You want five categories of numbers, so divide 1 by 5 and get 0.2. Every time `RAND()` gives a number between 0 and 0.2, you assign the first activity level, which is 1.2. The next "IF" means "otherwise" if the random number is less than 0.4, give the next activity level, 1.375. You can paste this into a cell and then "drag" it down to get lots of values. Notice that sometimes one value won't show up as much as the others—that's just the nature of randomness. Sometimes with a die you will roll five sixes in a row even though it isn't likely.

To generate the student grade levels, you can use the floor command, which rounds numbers down, as shown here:

```
=FLOOR(RAND() * (13-9) + 9,1).
```

Finally, to get the BMI, you can use the simpler random number command be discussed above, `RAND() * (30 – 18) + 18`. Figure 51 shows some simulated data.

Do you want to generate one population or some different groups?
How will you use the data provided by your simulation? This will determine whether you decide to generate a single high school or several different high schools. Let's generate a population for one high school and then think about how to use that process to simulate many high schools.

	C2				fx	=RAND()*(30–18)+18		
	A	B	C	D	E	F	G
1	Activity Levels	Grades	BMI				
2	1.55	12	20.5167959				
3	1.55	11	29.5277694				
4	1.2	10	19.4000179				
5	1.55	11	26.5424074				
6	1.375	12	29.8172257				
7	1.375	10	21.8424571				
8	1.375	10	22.0379307				
9	1.55	9	25.099647				
10	1.9	10	24.5541022				
11	1.55	10	20.9426602				
12	1.55	10	20.8703845				
13	1.55	12	25.2926758				
14	1.55	10	24.6530723				
15	1.375	9	21.0126314				
16	1.55	11	26.6724258				
17	1.375	11	26.1970576				
18	1.375	12	28.2000152				
19	1.2	9	29.9811486				
20	1.2	11	29.2533455				
21	1.7	12	20.4675876				

Figure 51: Excel output from the Lunch Crunch problem using the `RAND()` command three ways.

Imagine that the high school has 1000 students with equal numbers of students in each grade. You can assign each of the students a randomly generated BMI from above. A simple way to assign the students an activity level might be to generate a random number for each student between 1 and 5 and use that to assign the activity level. (We haven't specifically mentioned syntax for how to generate random *integer* values. Do a quick search now and see if you can figure out how to do it for your software of choice.)

Suppose instead you want to simulate a population of students where approximately the same number are sedentary as engage in extreme levels of activity. Suppose also that twice as many students have mild or heavy activity level, and three times as many students have moderate activity. Then you could generate random numbers from 1 to 9 and assign number 1 to sedentary; 2 and 3 to mild; 4, 5, and 6 to moderate; 7 and 8 to heavy; and 9 to extreme. You can use this idea to create your own tailor-made distributions.

A quick internet search can help you figure out how to generate numbers from all different kinds of distributions. While we have discussed Excel here, most software packages have the capability to generate random numbers. For example, MATLAB code written to simulate the caloric intake of high school students is included in Appendix B. We also provide guidance on how to write code with valuable comments so that you can look back at it later (or someone else look at it) and understand your thought process.

<table>
<tr><td>

5.2 EXAMPLE 2: FARMING DECISIONS

</td><td>

Approach 1: Optimization revisited with a program

Recall that in the farming problem from the Computations chapter, we wanted to determine the proportions of the crops a farmer could plant. One approach was to help the farmer maximize profit while also trying to meet a water restriction constraint, using the MATLAB solver `linprog.m`. Let's see how to incorporate this approach within a larger program. We saw there that a user would need to define several input vectors containing model and constraint coefficients (*f*, *A*, *b*, *Aeq*, *beq*, *lb*, *ub*) before calling the solver. If you wanted to change one of those and rerun the code, you could. But suppose you want to change the value of the water restriction limits over a range of say, twenty values; then you would need to manually change the parameter and then call the solver twenty times. Maybe that's not too bad, but you can imagine that it would be really annoying if you had to do it 200 times.

</td></tr>
</table>

Any time you find yourself doing the same computations repeatedly, or in this case making repeated calls to the same function, you should consider leveraging technology to do that work for you. In this case, we can give MATLAB directions about which parameter values we'd like it to use, and then tell it to repeat the computation by using a "loop."

Consider the case that a farmer has planted crops but then slowly the price of water begins to increase because of drought conditions. To see the impact on the farm profit, you could evaluate the profit model over a range of water price values and then plot the results. To outline this numerical experiment, you should first outline the steps you would want to take, before jumping in trying to write a program. This may be in the form of a numbered outline, a flow chart, or even *pseudocode:*

1. Set input for solving the optimization problem with linprog.m.
2. Call linprog.m to get the solution $X = [X_A, X_B, X_C]$. (Note: these lines of MATLAB code would look like what you entered into the workspace in the Computations chapter.)
3. Calculate the profit.
4. Examine the changes in profit as the price of water changes:
    (a) Define the values of water price to consider.
    (b) Loop over those values and calculate the cost of water based on the optimal crop distribution.
    (c) Calculate the adjusted profit using the new price of water and store those values.
5. Plot the profit as a function of the price of water.

The following lines of MATLAB code were saved in a file called farm_model.m and, when farm_model.m is called from the command prompt, those lines of code will be executed by the software.

```
%farming_code.m


f = [-4000,-3000,-8000];        % Coefficients in the profit model
                                % (S_A*Y_A), (S_B*Y_B), (S_C*Y_C)
A = [3,2,1];                    % Coefficients for water use, W_A,W_B,W_C
b = 235;                        % Water restriction limit W_R
Aeq = [1,1,1]                   % Coefficients so X_A+X_B+X_C <= 100
beq = 100
lb = [0,0,0];                   % Lower bounds on X_A, X_B, X_C
ub = [60,42,18];                % Upper bounds on X_A, X_B, X_C


[X, FVAL] = linprog(f,A,b,Aeq,beq,lb,ub)        % Call the solver


% Postprocessing, note: X_A=X(1), X_B=X(2), X_C=X(3)


profit = 4000*X(1)+3000*X(2)+8000*X(3); % Calculate the profit


PW = linspace(20,200,50);       % We will consider 50 values of the price
                                % of water between $20 and $200
for i = 1:length(PW)            % Loop over those values
  water = PW(i)*(3*X(1)+2*X(2)+X(3)); % Calculate the cost of water used
  P(i) = profit - water;   % Calculate the profit and store it in a vector
end


plot(PW,P);   % Plot the profit as a function of water prices
xlabel('Price of water');
ylabel('Profit')
```

We strongly suggest that, when possible, you do a few computations by hand (for example, here you could calculate the adjusted profit by calculating the cost of water for a few different water prices) and compare them to the software output to be sure you do not have any errors in your code. We demonstrate this principal on the Disease Model later in this chapter. Figure

52 shows the output from running this program and the plot that is generated. We suggest you play with the MATLAB Help menu to learn more about the built-in MATLAB functions `linspace`, `length`, `plot`, `xlabel`, and `ylabel`.

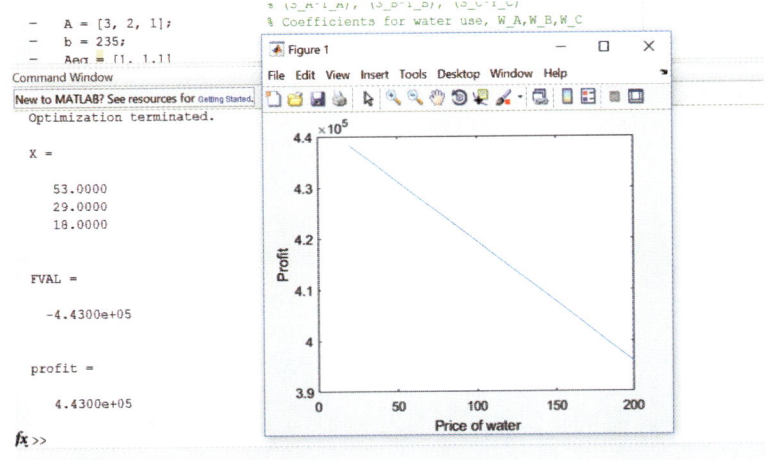

Figure 52: MATLAB output when solving the farm optimization problem and then varying the price of water, adjusting the profit model, and finally plotting the profit as a function of water price.

Approach 2: Write a virtual farmer simulator.

As mentioned in the Computations chapter, for the farm problem, it would be more realistic to allow the farmer to make a decision when land opens up rather than expecting her to replant the same crops over the entire two years. Since Crops A and B are four-month crops, they can be planted six times over two years. Also, the plants can go into the ground any month, so even though Crop C is an 8 month crop, it can still be planted any time except the last planting period, if we assume that we want everything out of the ground at the end of year two.

Figure 53 illustrates the possible decisions a farmer has when trying to plan out a two-year planting portfolio. A given year can be split up into three planting periods of four months per year. In the first planting periods, anything could be planted. After four months, the acres containing Crops A and B would be freed up, and anything could be planted there. Crop C would still be in the ground though. At the end of Period 2 in Year 1, Crop C would be harvested as well as whatever was planted for Crops A and B in Period 2, but the acres dedicated to the most recent planting of Crop C would still be off-limits for new crops. Any time land opens up, the farmer could make a decision to plant Crops A, B, or C given the amount of free land.

To this end, accumulated profit, water use, and proximity to consumer demand could be calculated every four months to help guide the farmer. For example, if in a four-month period the farmer didn't make much profit but used little water, he/she might be willing to plant a water intensive crop in the next planting period to get caught up financially. Figure 54 is a flow chart that shows how your code may be organized.

	Period 1: 4 Months	Period 2: 4 Months	Period 3: 4 Months
Year 1	Plant some amount of crops A, B, and C.	Crops A and B are harvested, Crop C is still in the ground, can plant Crops A, B, or C or leave land empty.	Crops A, B, and the original C are harvested, Crop C that was just planted is still in the ground, can plant Crops A, B, or C or leave empty.
Year 2	Crops A, B, and the previous C are harvested, Crop C that was just planted is still in the ground, can plant Crops A, B, or C or leave empty.	Crops A, B, and the previous C are harvested, Crop C that was just planted is still in the ground, can plant Crops A, B, or C or leave land empty.	Crops A, B, and the previous C are harvested, Crop C that was just planted is still in the ground, can plant Crops A or B or leave land empty.

Figure 53: Possible planting opportunities for our farmer.

This sort of model is time dependent with dynamic decisions and is a great example of how computer programing can be used to build an interactive software program. Any software tool that allows for computer programing (MATLAB, Python, Mathematica, etc.) that would prompt the user to enter a value could be used here. Instead of providing syntax that is tied to any particular computing language, we outline how this approach to the farming model could be turned into a simulator acting as a virtual farmer.

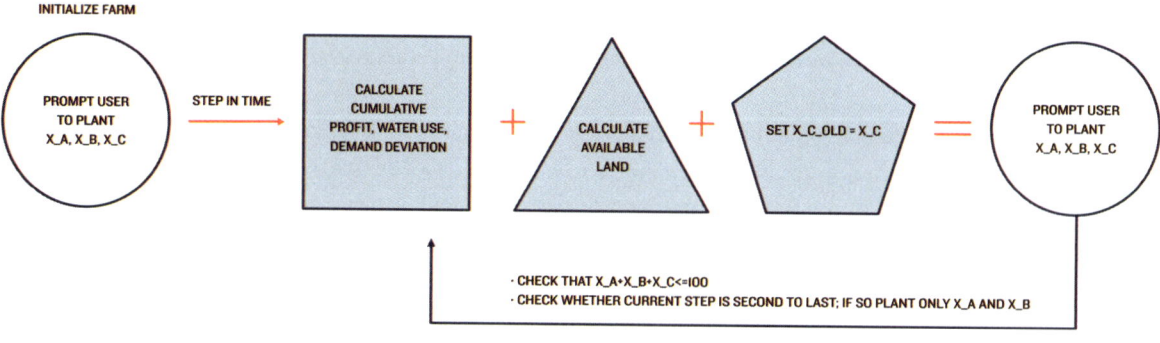

Figure 54: Virtual farming simulation algorithm flow chart to help you organize the code structure first.

5.3 EXAMPLE 3: DISEASE SPREAD

In the Computation chapter, we outlined several ways you might use technology in finding a solution for the Disease Spread question (glance in the Appendix if you need to review the question). Approaches 2 and 3 in that chapter worked toward solving the differential equation model of the form

$$\frac{dI}{dt} = kI(t)\big(P - I(t)\big),$$

where I is the number of infected individuals, P is the total population, and k is a positive constant.

If you want to use this differential equation model because you believe it best describes the physical scenario, but you don't feel comfortable using either a symbolic computation tool or an out-of-the-box solver provided in some programming software (as shown in Approaches 2 and 3 in the Computation chapter), you can still find a numerical solution by doing a simulation in a spreadsheet. Let's discuss how that would work, using the parameter values $k = 0.0006$, $P = 1000$, and $I_0 = 20$. That is, we are solving

$$\frac{dI}{dt} = 0.0006I(t)\big(1000 - I(t)\big),$$

under the assumption that 20 people in the population are initially infected with the disease.

The process we will use here is called the Forward Euler method, which takes advantage of the fact that the derivative is the same thing as the slope of the tangent line. This technique (a) can be used to solve most first order differential equations, (b) can be adapted to solve other types of differential equations, and (c) is the basis of more sophisticated numerical techniques, so it's a nice technique to investigate.

We'll demonstrate the method on this example, but we don't want to overwhelm you with too many details. You may want to do an internet search for Forward Euler if you'd like to have more details about how the method works and some considerations when using it. The focus here is how you can use technology to simulate how the infected population changes over time.

We start with the initial population, $I_0 = 20$. That is, when $t = 0$ days, $I = 20$. In other words, we know that the point $(0, 20)$ is on the graph of the solution. We also know what the slope of the solution curve is at that point, because we have an equation for dI/dt:

$$\left. \frac{dI}{dt} \right|_{t=0} = kI(0)\big(P - I(0)\big)$$

$$= kI_0\big(P - I_0\big)$$
$$= (0.0006) \cdot 20 \cdot (1000 - 20)$$
$$= 11.76.$$

Now we can write the equation of the line through the point $(0, 20)$ with slope 11.76, as follows (recall that the independent variable is t and the dependent variable is I):

$$I - 20 = 11.76(t - 0),$$
$$I = 11.76(t - 0) + 20.$$

Hence, we have a line with slope 11.76 and y-intercept (or I-intercept, in this case) 20.

We can use this linear function as an approximation for the true solution. We assume that the solution to the differential equation is approximately the same as this line for points nearby. Perhaps we'll assume it's a good enough approximation through time $t = 1$. We can find the number of infected individuals at $t = 1$:

(9)

$$I(1) = 11.76(1 - 0) + 20 = 31.76.$$

[Note: While it's not possible to have a fractional person (or, more specifically, 0.76 of an infected individual), that doesn't deter us from continuing to approximate the solution using this method. We do suggest noting that something unrealistic has occurred, and encourage you to re-examine it later when assessing your model.]

Thus we have a line segment from (0, 20) to (1, 31.76).

Now you can imagine us starting the process over. In other words, we assume that the point (1, 31.76) is on the solution curve, and we can use the derivative to give us the slope at that point:

$$\frac{dI}{dt}\bigg|_{t=1} = kI(1)\big(P - I(1)\big)$$

$$= (0.0006) \cdot 31.76 \cdot (1000 - 31.76)$$
$$= 18.45.$$

As before, we can find the equation of the line through the point (1, 31.76) with slope 18.45:

$$I - 31.76 = 18.45(t - 1),$$
$$I = 18.45(t - 1) + 31.76.$$

We will assume that this makes a good enough approximation for the solution through $t = 2$. So we estimate the population at time $t = 2$ to be

$$I(2) = 18.45(2 - 1) + 31.76 = 50.21.$$

Figure 55 shows this process visually.

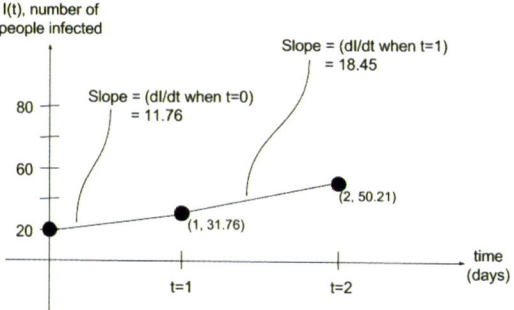

Figure 55: Visualization of the Forward Euler method in solving the differential equation $dI/dt = 0.006I(t)(1000 - I(t))$ with initial condition $I(0) = 20$, where $I(t)$ is the number of people infected by a disease at time t.

Now that we have done a couple of iterations by hand, we can see that we are just repeating the same calculations over and over again. Rather than continue by hand, we can transition to performing the computations in a spreadsheet.

Figure 56 shows how you might implement the Forward Euler method in a spreadsheet. As we noted in the Computation chapter, it's advantageous to put parameter values in their own cells so that they can be varied during a sensitivity analysis, so that's why the k and P values appear in cells B6 and B7, respectively. Once you've determined how you're going to lay things out in your spreadsheet, you can start entering the starting values. For instance, since we are determining the population every day, we'll start by creating the column containing $0, 1, 2, \ldots,$ 6. (Note: can you use the "drag" feature to do this quickly?)

B16		fx	=B15+C15*(A16–A15)	
	A	B	C	D
1	Number of people infected with a disease			
2				
3	model:	dI/dt=k*I*(1000-I)		
4				
5	parameter values:			
6	k =	0.0006		
7	P =	1000		
8				
9	initial number of infected people:			
10	I_0 =	20		
11				
12				
13	time (days)	number infected (I)	rate dI/dt	
14	0	20	11.76	
15	1	31.76	18.45078144	
16	2	50.21078144	28.61379532	
17	3	78.82457676	43.56675772	
18	4	122.3913345	64.44701743	
19	5	186.8383519	91.1578693	
20	6	277.9962212	120.4285933	
21	7	398.4248145	143.809489	
22				

Figure 56: Using a spreadsheet to implement the Forward Euler method for the disease spread differential equation model.

Next we enter the value for the initial number of people infected into cell B14. Rather than enter 20, though, we will type

$$=B10,$$

so that if we later want to vary the initial number of people infected, we can make that change in just one place, cell B10, and the rest of the spreadsheet will update to reflect that change.

We then move to cell C14, where we want to have the value of the dI/dt when $t = 0$. We'll use the formula $dI/dt = kI(t)(P - I(t))$, so in cell C14 we type

$$=\$B\$6*B14*(\$B\$7 - B14).$$

The dollar signs are used so that when you "drag" the formula, you'll always refer back to the k and P you defined at the top of the spreadsheet.

Next we need to use the rate we found at $t = 0$ to estimate the number infected at $t = 1$ using (9). We achieve this by typing

$$=B14+C14*(A15 - A14)$$

into cell B15.

Finally, now that we have formulas in columns B and C, we can simply "drag" down those formulas to get the spreadsheet shown in Figure 56.

It's a good sign that the numbers we computed by hand appear in the spreadsheet! However, you should always be mindful of the reasonableness of your output. Continually asking yourself "does this make sense?" will help you continue to move forward as you iterate through the modeling process. In this example, we have modeled a disease that's continuing to infect more and more people each day—nearly 400 people have been infected by the 7th day! But, because the overall population is 1,000 and we have not allowed for any additions to (or subtractions from) the population, we know that the diseased population will not exceed 1,000. As a result, we should be prepared to observe a decrease in the *rate* at which the disease spreads. With the work you've done on this problem, you are now in a position to see whether your computational model is consistent with your real-world intuition; continue to drag down on your spreadsheet to find when the infection rate drops below 1 person per day. Additionally, consider graphing the number of infected individuals over time to identify the long-term behavior of the solution; in this case, an asymptote at 1,000.

Now that you have verified the reasonableness of your computational result, you can make updates and change values with ease. Using the disease model as an example, what if the number of infected people is 25? Or 5? More generally, how does a small change in a parameter value alter the time required for the population to become infected? Alternatively, what if you are interested in tracking the spread of the disease over shorter (or longer) time intervals? Does this improve (or worsen) your computational result(s)? For example, what if you are interested in tracking the disease spread at 6 hour (0.25 day) intervals? Is Euler's method still appropriate? Does it give you a better (or worse) solution?

Building a simulation is a great way to investigate complicated modeling problems. Maximize the value of this tool by altering values (in a meaningful way) and continuing to keep track of the "reasonableness" of your results.

5.4 TAKE-AWAY MESSAGES

- When computations may become tedious because you have to repeat them for different parameters or times, you may want to create a simulation.
- Simulations are often used when real experiments would be impossible, dangerous, or too costly to perform, or when you want to simulate data because none is unavailable to you.
- Simulation can be used to generate data to plug into and test your model. You may also want to try some extreme values to see when and if your model breaks (or doesn't provide a useful solution).
- It can be helpful to specify an underlying distribution (such as uniform, normal, or bimodal) for your data as well as characteristics such as mean and standard deviation.
- Random number generators can also be very helpful in simulating data. If you generate data from a function, this could be a way to introduce some more realistic messiness and scatter.
- You can use programming to generate data with specific values assigned for the parameters in the model.
- Pseudocode or algorithm flow charts are outlines of how you want your computer program or simulation code to work. It doesn't have to be in any special language, but it is a great first

step to help you organize the different processes that you are trying to put together into a working program.

- It may be helpful to *do a few iterations of computations by hand* before jumping into using technology. Sometimes it's difficult to try to think of the appropriate formula to use while simultaneously figuring out how to make your technology work. Doing a couple of iterations by hand allows you to solidify your understanding of the formula first, before you even think about the software implementation. You can also compare your pencil-and-paper computations with the ones your technology provides to confirm that the software is producing the output you think it should be.
- It's always good practice to implement a simpler model successfully and then revise and adapt it to make a more sophisticated version.
- Writing useful code takes practice. Check out Appendix B to see an example of the Lunch Crunch simulation using MATLAB with three different coding styles.

Concluding thoughts on programming and simulation

Learning a programming language is a marketable skill in the long run, but it may take time, commitment, and patience. Luckily there are lots of resources available (more on this in the next chapter). In the context of this book, designing a simulation or a specially tailored algorithm can be an innovative way to approach a modeling scenario. In the end, you may develop a powerful mathematical tool that is able to address more realistic problems because you took the initiative to try something new.

RESOURCES: CHOOSING YOUR TECH TOOL

In this section we provide some details about the tools we use throughout this guide and highlight their capabilities. If you see one that sounds interesting, you can go their website and start to explore. All of these tools are well supported and used in the real-world to solve a wide range of complex problems. They are all equipped with helpful, user-friendly manuals, tutorials, and often even videos to help get you started or to teach you how to use special features.

6.1 SPREADSHEETS

Spreadsheets are one of the most popular computational tools used in schools and in industry. (Engineers and financial planners use spreadsheets on a daily basis!) The technical definition of a spreadsheet is "an electronic document in which data is arranged in the rows and columns of a grid and can be manipulated and used in calculations." The entries in a spreadsheet are referred to as cells that are labeled according the their position in the spreadsheet (row and column location). When you find data or are provided with it you may find that it's in a file prepackaged as a spreadsheet.

Spreadsheets allow you to organize and format large data sets for easy manipulation and analysis. The cells in a spreadsheet can hold numbers (in a variety of formats), text (strings), Boolean values, or formulas. The power of a spreadsheet is that calculations can be done over entire rows or columns at once. Spreadsheet programs come with a range of options for statistical analyses (averages, maximum values, etc.) and can also be used to create graphs (scatter plots, bar charts, etc.).

What are some examples of spreadsheet software tools?

• Microsoft Excel (Excel) is probably one of the most commonly used spreadsheet software programs. Although it costs money, it is a feature of Microsoft Office that can be found on many computers (and is likely available in most school and university computer labs).

• Apache Open Office Calc can be downloaded for free at *https://www.openoffice.org*.

• Google Spreadsheets at *https://www.google.com* can be used with any (free) Google account.

Spreadsheet tips

• When you try to download a data set, it may be available to you as a .csv file. This is a *comma-separated values* file, which stores data (numbers and text) as plain text.

• Many spreadsheets have additional "add-ins" that are readily available to download and will automatically install with your software package with little effort. For example, we use the Excel Solver add-in to approach one of the models for the farming problem in the Computations chapter.

• In the Programming and Simulation chapter, we show how you can use a spreadsheet to generate data. This data can be used to test your models or demonstrate how they work in different situations.

• The Visualization chapter will provide some tips about how to create charts and graphs from the data in a spreadsheet.

6.2 MATHEMATICAL SOFTWARE

There are countless mathematical software packages at your fingertips that could be used to advance your math model! As you explore and use different ones, you'll likely have a favorite that is your go-to tool, and you will quickly become an expert in using it.

Each one has unique features, but most have some common capabilities that include built-in functions that any calculator would have. Their strengths lie in additional solving capabilities to tackle more complex problems, like finding the roots of a function or finding the maximum of a complicated function of several variables. Software packages also have built-in visualization and statistical tools to help you analyze your results.

What are some examples of mathematical software packages?

We mention some commonly used software here, but be aware that the list of mathematical tools is constantly growing and changing, so you shouldn't be restricted to what you see here.

- **MATLAB** is a licensed computing environment developed and supported by Mathworks that is widely used in universities and industry to solve real world problems. It can be used as a straightforward mathematical tool much like a graphing calculator but also has powerful built-in scientific computing tools, and the MATLAB language can be used to create specialized computer code. You can get started with the language using an online tutorial. You don't have to have any programming experience, but if you do, you will be able to use some of those same ideas in MATLAB. The MAT stands for Matrix, so it can help to have some experience with matrices. MATLAB has an easy-to-use graphical interface, and learning the language is made easier because it has a detailed Help menu and an online support community. Its functionality includes computations that use an extensive built-in function library, ability to solve large problems (matrix systems, nonlinear equations, differential equations, optimization problems, etc.), algorithm development, simulation, visualization, and statistical analysis. It also has functionality with other software platforms (for example, MATLAB could call C++ code or read from a spreadsheet). The workspace is shown in Figure 57. *https://www.mathworks.com/*

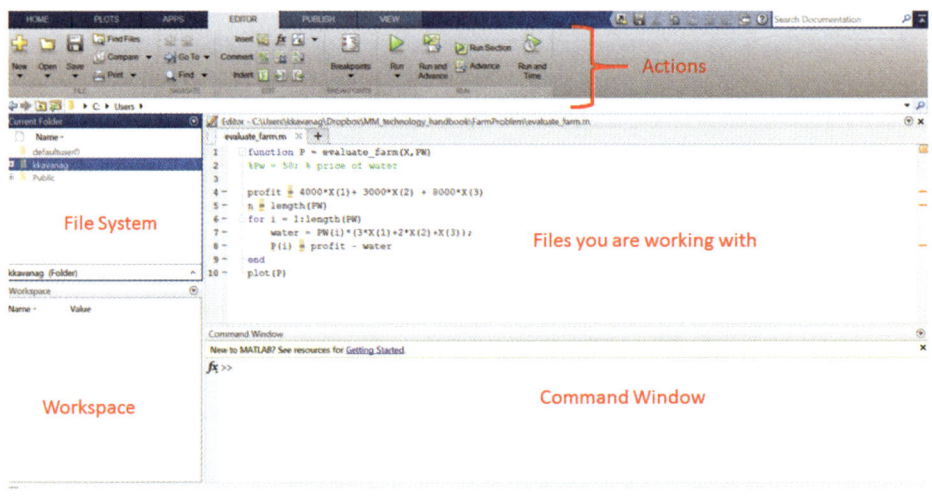

Figure 57: You can type scripts in the Command Window just like you would a regular calculator, or you can call built-in MATLAB solvers or functions that you wrote (which appear as a tab above the Command Window). In the workspace, you would see which variables you are working with and can click on them to see what value they hold.

- **Mathematica** was created as a computer algebra program. It tends to be used for symbolic computation, where you might want to do something like solve a polynomial or take a derivative. Mathematica also has a great deal of built-in mathematical, scientific, and socio-economic information. It has its own programming language that can be used to create programming scripts in files called notebooks. In addition, Mathematica has some capability for "natural language input," which means you can ask it to do things with instructions that are more like spoken sentences than typical programming instructions. *https://www.wolfram.com/mathematica/*

- **Wolfram Alpha** is a free online "answer engine" that performs computations and mathematical tasks. It understands commands that are more like spoken sentences, provides multiple solution strategies and types, gives algebraic and graphical versions of the solutions, and is capable of doing symbolic algebra, calculus, or stats. It can also be used on phones as well as computers or tablets. *https://www.wolframalpha.com/*

• **R** is a statistical software environment, considered to be one of the most comprehensive statistical software tools available today. R can handle large data sets and is equipped with analytical capabilities from the fairly basic to the cutting-edge. For example, it can be used to generate regressions or simulate data to use in your model when you cannot find any. It has a variety of graphing capabilities that can be useful in cleaning up or taking an initial assessment of your data. It may take some time to become confident with R and to really understand the powerful underlying tools, but it would be worth it to start early and take time to slowly learn all of its features. *https://www.r-project.org/*

• **Scilab, Octave, and SageMath** are free and open-source computing environments with some of the same features and toolboxes as MATLAB or Mathematica.

Mathematical software tips

• If you have a hard time reading through the user-manual documentation, do not hesitate to look online for examples, message boards, or other tutorials that might help you move forward. Just getting the right keywords in your search can save time.

• Each of these software tools will have its own environment where you will execute mathematical commands, write subroutines or functions, generate plots, and basically do your work.

• When getting started using software, if you have a question or are getting an error message, try typing or copying it into a search engine to see if there are suggestions about how other people resolved the same issue.

6.3 PROGRAMMING LANGUAGES

Learning to code is exciting and opens the door to many problem solving possibilities. You can also develop marketable job skills for your future! As you learn to code in a programming language you will become familiar with its structure and use it to create software programs, scripts, or other sets of instructions for computers to execute. There are many languages to choose from, which can be intimidating at first, but with practice, everyone can learn to code!

To use most program languages, you will likely need to install a package (a set of files) which may or may not come with a graphical user interface (GUI). In addition, certain features require additional "libraries" in order to operate. A *library* is a set of files with prewritten code that you can use so you do not have to start from scratch. Most languages have a special math library that contains built-in functions that go beyond basic arithmetic operations. Look up some libraries for the language you choose.

A *compiled language* is a programming language that requires additional software called a compiler that translates your code into a new format that the computer can execute. Some examples are C++, Java, FORTRAN, and Visual Basic. An *interpreted language* is a programming language in which an internal "interpreter" runs your computer code without the additional step of compiling. You can use basic scripts or commands to run your code.

Python is an example of an interpreted code.

Python has quickly grown to become one of the most widely used programming languages and continues to be developed and improved. It was designed with succinct commands in line, so that one line of code in python does the work of several lines of code in other languages. It is considered fairly easy to learn and a stepping-stone to learning other object-oriented languages such as Java or C++. It may be a good choice for first-time programmers or as a new tool for experienced programmers to pick up in a short time. See *https://www.python.org/*.

A good resource for getting started with a wide range of languages is `Codeacademy`, an online learning tool where you can work at your own pace through interactive tutorials and level-up as you go.

Programming language tips

• Learning programming is more than learning a language. You are also learning new technology, best practices, and a way of thinking such that you break down steps into an algorithm. This is not easy. If it was, we would all be excellent computer scientists! So, take your time and *read over all the documentation and tutorials you can,* and you'll get the hang of it.

• Likely the programming language you choose will come with a framework for writing code. You will usually use a special text editor rather than a writing program like Microsoft Word. The special text editors sometimes color-code the commands in a file as one way to help you stay organized.

• Before you write any computer code, you should start with an informal outline of what you want your program to do, often called a *pseudocode*. It doesn't contain any syntax, just the basic steps you take to execute the purpose of your program.

• One of the best habits you can develop is to include detailed comments while you are writing your code that explain what your code is doing on each line. You will be surprised how much you can forget when you stop working on a project for a few days, and then you go back to read your code. It's completely common to have no idea what you were doing! Comments will save you lots of time in backtracking and make it easier to share your code with others so they can understand what you were doing. Check out an example on the Lunch Crunch problem in Appendix B.

• Another valuable habit is to name your variables appropriately so that when you go back to look at your code, it matches the notation or names in your math model.

6.4 CONCLUDING THOUGHTS ON USING SOFTWARE

Coding is an incredibly important skill because it can enhance your ability to perform many different kinds of tasks at school, work, and home. We believe that everyone can be a successful user of computing languages, and mathematical modeling is a great way to get started. We hope this guide will help you choose a tool that works well for the job you want it to accomplish and use it effectively. If you are just getting started, stick with it and practice until it gets easier. Many times working with a partner can make coding more fun and help you generate more ideas about how to choose the right tool and write clear, efficient, useful code.

APPENDIX A: PRIMARY EXAMPLES USED THROUGHOUT THIS GUIDE

We demonstrate the use of technology within the modeling process by looking at several modeling questions in detail. We state those problems directly below, and then explore them more throughout the remainder of this guide. We should note that we will also use some other examples to highlight specific ideas, but these four are consistently discussed throughout.

I. Lunch Crunch: Can Nutritious Be Affordable and Delicious? [1]

First Lady Michelle Obama spearheaded an initiative on good nutrition that led to passage of the Healthy, Hunger-Free Kids Act of 2010. Implementation of the act, however, revealed the competing preferences of the school lunch program's three major stakeholders. Students care most about taste and quantity; school districts are concerned about affordability; and the federal government, which provides financial support, wants to promote lifelong healthy eating habits.

Schools have seen the cost of offering lunch go up (since healthier foods are often more expensive), while participation goes down (students are less satisfied with school lunch, either because it doesn't taste as good or it isn't filling enough), causing a fiscal crisis for some school districts.[5]

The USDA has asked your consulting firm to provide a report with mathematically founded insights into the problem; you should address at minimum the following three concerns:

1. **You are what you eat?** Students' caloric needs at lunch depend on how active they are, whether they eat breakfast, and a host of other factors. Develop a mathematical model that takes as input a student's individual attributes, and outputs the number of calories that a student with those attributes should eat at lunch.

2. **One size doesn't necessarily fit all.** The guidelines dictated by the Healthy, Hunger-Free Kids Act of 2010 are based on meeting the needs of an "average student."[6] However, meeting the average need may not necessarily provide the right amount for many students. Now that you've identified attributes that affect caloric needs at lunch, create a model to determine the distribution of U.S. high school students among each of these categories. If every student eats the standard school lunch, what percentage of students will have their caloric needs met at lunch?

3. **There's no such thing as a free lunch.** A sample school district has a weekly budget of $6 per student for the purchase of food only. Leverage math modeling to develop a lunch plan (using food categories) that stays within the budget, meets the nutritional standards, and appeals to students. What changes would you make if your budget was increased by $1?

You may want to take into account how your model could be applied across different geographic and/or socio-economic regions. Your report to the USDA should include a one-page summary of your findings.

You may find the following websites helpful:

http://www.globalrph.com/estimated_energy_requirement.htm
http://www.cdc.gov/mmwr/pdf/ss/ss5905.pdf
http://www.cdc.gov/growthcharts/charts.htm#Set3
http://www.amstat.org/censusatschool/about.cfm

II. Outbreak? Epidemic? Pandemic? Panic?

We all dread getting sick. Years ago, illness didn't spread very quickly because travel was difficult and expensive. Now thousands of people travel via trains and planes across the globe for work and vacation every day. Illnesses that were once confined to small regions of the world can now spread quickly as a result of one infected individual who travels internationally. The National Institutes of Health and the Center for Disease Control are interested in knowing how significant the outbreak of illness will be in the coming year in the U.S.

III. Farming Example: Water you going to plant? Lettuce help you!

Efficient water use is becoming increasingly important as periods of sustained drought and overuse of aquifers put available resources in jeopardy. In regions of intense agriculture, this problem requires immediate attention to ensure food security and a stable economy. Balancing profit, water use, and economic demand requires deciding what distribution of crops to plant under uncertain climate conditions and changing water policies. For example, some crops which are profitable may require significant irrigation. It is not always the case that crops which are high in demand are the most profitable either. In addition, farmers may plant an entire farm and, depending on the growing cycle and effort required, be locked in to a particular crop portfolio for a few years (for example, raspberries have a two-year growing cycle). During that time frame, the price of water may increase drastically while rain events are scarce.

Consider a 100-acre farm and three possible crops to plant. You can choose from Crop A, which is high in demand and profitable, but water intensive; Crop C, which has the lowest demand and lowest profit but uses the least amount of water; and Crop B, which will be in the middle for all three properties. Create a mathematical model to determine what to plant over a two-year time horizon. Assume the following crop properties: We assume that, to meet the current demand, this farm typically allocates 50% of its acreage to Crop A, 35% to Crop B, and 15% to Crop C. You may want to consider that crops have different growing cycles. For example, consider that Crops A and B are in the ground for four months before harvesting, while Crop C must be in the ground for eight months. See Table 6.

Crop	A	B	C
Yield (boxes/acre)	1000	1000	4000
Sales price ($/box)	4.00	3.00	2.00
Water use (acre-ft/acre/year)	3	2	1

Table 6: Model parameters for each crop.

IV. From Sea to Shining Sea: Looking Ahead with the National Park Service [2]

The National Park system of the United States comprises 417 official units covering more than 84 million acres. The 100-year old U.S. National Park Service (NPS) is the federal bureau within the Department of the Interior responsible for managing, protecting, and maintaining all units within the National Park system, including national parks, monuments, seashores, and other historical sites.

Global change factors such as climate are likely to affect both park resources and visitor experience [1] and, as a result, the NPS's mission to preserve unimpaired the natural and cultural resources and values of the National Park system for the enjoyment, education, and inspiration of this and future generations. Your team can provide insight and help strategize with the NPS as it starts its second century of stewardship of our nations park system.

1. **Tides of change.** Build a mathematical model to determine a sea level change risk rating of high, medium, or low for each of the five parks below for the next 10, 20, and 50 years.

Acadia National Park, Maine
Cape Hatteras National Seashore, North Carolina
Kenai Fjords National Park, Alaska
Olympic National Park, Washington
Padre Island National Seashore, Texas

You may use provided data[7] on sea level to build the model. Explain your interpretation of high, medium, and low. Could your model realistically predict those levels for the next 100 years?

2. **The coast is clear?** In addition to the phenomena listed above, the NPS is investigating the effects of all climate-related events on coastal park units. Develop a mathematical model that is capable of assigning a single climate vulnerability score to any NPS coastal unit. Your model should take into account both the likelihood and severity of climate-related events occurring in the park within the next 50 years. Some or all of the provided data may be used to assign scores to the five national park units identified in question 1.

3. **Let nature take its course.** The NPS works to achieve its mission with limited financial resources that may vary from year to year. In the event that costs—such as those caused by climate-related events—exceed revenues and funding, NPS must prioritize where to spend monies.

Consider incorporating visitor statistics and your vulnerability scores (and possibly other variables that may be considered priorities) to create a new model that predicts long-term changes in visitors for each park. Use this output to advise the NPS about where future financial resources should go.

Links to help get started:
- *http://recode.net/2016/01/04/gm-invests-500-million-in-lyft-and-strikes-strategic-autonomous-car-alliance/*
- *http://www.businessinsider.com/r-ford-rd-chief-says-automaker-wants-to-develop-ride-hailing-services-2015-12*
- *http://time.com/4188430/general-motors-maven-car-sharing/*
- *http://www.greenbiz.com/article/zipcar-google-and-why-carsharing-wars-are-just-beginning*
- *http://nhts.ornl.gov*

ENDNOTES

[5] *http://healthland.time.com/2013/08/29/why-some-schools-are-saying-no-thanks-to-the-school-lunch-program/*

[6] *http://cspinet.org/new/pdf/new-school-meals-faq.pdf*

[7] *https://m3challenge.siam.org/node/336*

APPENDIX B: USING COMMENTS — BEST PRACTICES WHEN WRITING CODE

The following are three examples of MATLAB code that achieve the *exact same thing*—simulating a group of high school students and determining how many calories they are likely to expend in a given day (for possible use in a solution to the Lunch Crunch problem). Even though MATLAB will do the same computations using all three, the experience when you try to use or edit the code is very different for each.

The first code is difficult to read. It's not easy to tell what the variables mean, and there's no easy way to know what the code hopes to accomplish.

The second code is a little better. The variable names are more meaningful; while it might be a little more annoying to type out because the names are longer, it's worth it to give meaningful variable names so your code is understandable to others.

The second code also has *comments*. Comments appear after a % sign (in MATLAB) and are notes that are meant to help the person reading the code know what each section or line is supposed to do. MATLAB doesn't "look" at anything on a line to run it after that percent sign. Unfortunately, while the second code has comments, they do not, in general, help a reader understand anything more about how the code works.

Compare that to the third code, which we would describe as a well-commented code. In this code, the comments provide insight into why the code appears as it does. This is extremely helpful for someone who might try to use this code (and perhaps even for the author, in looking back at the code).

CODE 1: WITHOUT COMMENTS

```
N = 20;
mf = randi([0 1],N,1);
g = randi([9, 12],N,1);
a = g + 5.5;
ALF = rand([N,1]);
for i = 1:N
    if ALF(i)<0.2
        ALF(i) = 1.2;
    elseif ALF(i)<0.4
        ALF(i) = 1.375;
    elseif ALF(i)<0.6
        ALF(i) = 1.55;
    elseif ALF(i)<0.8
        ALF(i) = 1.7;
    else
        ALF(i) = 1.9;
    end
end
W = zeros(N, 1);
H = zeros(N, 1);
BMI = zeros(N, 1);

for i = 1:N
    if mf(i) == 1
        if g(i) == 9
            H(i) = (64.5 + 1.5*randn(1,1))*2.54;
            BMI(i) = 19 + randn(1,1);
        elseif g(i) == 10
            H(i) = (66.5 + 1.5*randn(1,1))*2.54;
            BMI(i) = 20 + randn(1,1);
        elseif g(i) == 11
            H(i) = (68.5 + 1.5*randn(1,1))*2.54;
            BMI(i) = 21 + randn(1,1);
        else
            H(i) = (69 + 1.5*randn(1,1))*2.54;
            BMI(i) = 21.5 + randn(1,1);
        end
        W(i) = (BMI(i) * H(i)^2)/10000;
        BMR = 88.362 + 13.397*W(i) + 4.799*H(i) - 5.677*a(i);
    else
        if g(i) == 9
            H(i) = (63.5 + 1.5*randn(1,1))*2.54;
            BMI(i) = 19.5 + 1.5*randn(1,1);
        elseif g(i) == 10
            H(i) = (64 + 1.5*randn(1,1))*2.54;
            BMI(i) = 20 + 1.5*randn(1,1);
        elseif g(i) == 11
            H(i) = (64 + 1.5*randn(1,1))*2.54;
            BMI(i) = 20.5 + 1.5*randn(1,1);
        else
            H(i) = (64 + 1.5*randn(1,1))*2.54;
            BMI(i) = 21 + 2*randn(1,1);
        end
        W(i) = (BMI(i) * H(i)^2)/10000;
        BMR = 447.593 + 9.247*W(i) + 3.098*H(i) - 4.330*a(i);
    end
end

C = 1.1*BMR.*ALF
```

CODE 2: SOME COMMENTS

```
num_students = 20; %Set the number of students equal to 20
is_male = randi([0 1],num_students,1); %Make a vector of ones and zeros to
grade = randi([9, 12],num_students,1); %Vector of numbers from 9 to 12
age = grade + 5.5; %add 5.5 to grade to get age
ALF = rand([num_students,1]); %random numbers between 0 and 1
for i = 1:num_students
    if ALF(i)<0.2
        ALF(i) = 1.2; %sedentary
    elseif ALF(i)<0.4
        ALF(i) = 1.375; %mild
    elseif ALF(i)<0.6
        ALF(i) = 1.55; %moderate
    elseif ALF(i)<0.8
        ALF(i) = 1.7; %heavy
    else
        ALF(i) = 1.9; %extreme
    end
end

weight_in_kg = zeros(num_students, 1); %vector of zeros for weight
height_in_cm = zeros(num_students, 1); %vector of zeros for height
BMI = zeros(num_students, 1); %vector of zeros for BMI
for i = 1:num_students
    if is_male(i) == 1 %if the entry in is male happens to be 1...
        if grade(i) == 9 % if student is in grade 9...
            height_in_cm(i) = (64.5 + 1.5*randn(1,1))*2.54; %get height
            BMI(i) = 19 + randn(1,1); %get BMI
        elseif grade(i) == 10 % if student is in grade 10...
            height_in_cm(i) = (66.5 + 1.5*randn(1,1))*2.54; %get height
            BMI(i) = 20 + randn(1,1); %get BMI
        elseif grade(i) == 11 % if student is in grade 11...
            height_in_cm(i) = (68.5 + 1.5*randn(1,1))*2.54; %get height
            BMI(i) = 21 + randn(1,1); %get BMI
        else %otherwise...
            height_in_cm(i) = (69 + 1.5*randn(1,1))*2.54; %get height
            BMI(i) = 21.5 + randn(1,1); %get BMI
        end
        weight_in_kg(i) = (BMI(i) * height_in_cm(i)^2)/10000;
            %weight equals (BMI * height^2)/10000
        BMR = 88.362 + 13.397*weight_in_kg(i) + 4.799*height_in_cm(i) ...
            - 5.677*age(i); %BMR depends on weight, height and age
    else %otherwise...
        if grade(i) == 9 % if student is in grade 9...
            height_in_cm(i) = (63.5 + 1.5*randn(1,1))*2.54; %get height
            BMI(i) = 19.5 + 1.5*randn(1,1); %get BMI
        elseif grade(i) == 10 % if student is in grade 9...
            height_in_cm(i) = (64 + 1.5*randn(1,1))*2.54; %get height
            BMI(i) = 20 + 1.5*randn(1,1); %get BMI
        elseif grade(i) == 11 % if student is in grade 9...
            height_in_cm(i) = (64 + 1.5*randn(1,1))*2.54; %get height
            BMI(i) = 20.5 + 1.5*randn(1,1); %get BMI
        else %otherwise...
            height_in_cm(i) = (64 + 1.5*randn(1,1))*2.54; %get height
            BMI(i) = 21 + 2*randn(1,1); %get BMI
        end
        weight_in_kg(i) = (BMI(i) * height_in_cm(i)^2)/10000;
            %weight equals (BMI * height^2)/10000
        BMR = 447.593 + 9.247*weight_in_kg(i) + 3.098*height_in_cm(i) ...
            - 4.330*age(i); %BMR depends on weight, height and age
    end
end

calories_expended = 1.1*BMR.*ALF %multiply 1.1 times BMR times ALF
```

CODE 3: WELL-COMMENTED

```matlab
%%%%%%%%%%%%%%%%%%%%%%%%%%%%%%%%%%%%%%%%%%%%%%%%%%%%%%%%%%%%%%%%%%%%%%%%%%%%
% Generate sample students in a high school and then compute the daily
% calories expended by the students.
%%%%%%%%%%%%%%%%%%%%%%%%%%%%%%%%%%%%%%%%%%%%%%%%%%%%%%%%%%%%%%%%%%%%%%%%%%%%
%User can change the number of students in the high school:
    num_students = 20;
%Male or female: randomly assign to each student a gender
%                                       (here 1 indicates student is male)
    is_male = randi([0 1],num_students,1);
%Grade level: randomly assign a grade level, 9 through 12, to each student
    grade = randi([9, 12],num_students,1);
%Set corresponding age for students: we assume every 9th grader is 14.5,
%every 10th grader is 15.5, every 11th grader is 16.5, and every 12th
%grader is 17.5 years old. In other words, if we add 5.5 to grade, we get
%the student's corresponding age.
    age = grade + 5.5;
%Activity Level Factor (ALF): each student is assigned to one of 5 different
%categories (from sedentary to extremely active)
    ALF = rand([num_students,1]);           %This assigns a random number between 0
                                            % and 1 from a uniform distribution.
                                            % We split the interval in 5 pieces and
                                            % assign a category to each
                                            % subinterval. Then the ALF can
                                            % assigned based on the category, as
                                            % follows:
    for i = 1:num_students
        if ALF(i)<0.2
            ALF(i) = 1.2; %sedentary
        elseif ALF(i)<0.4
            ALF(i) = 1.375; %mild
        elseif ALF(i)<0.6
            ALF(i) = 1.55; %moderate
        elseif ALF(i)<0.8
            ALF(i) = 1.7; %heavy
        else
            ALF(i) = 1.9; %extreme
        end
    end
%For weight, height, BMI, and BMR:
    % We pre allocate space in the following vectors for heights, weights,
    % and BMI of each students
    weight_in_kg = zeros(num_students, 1);
    height_in_cm = zeros(num_students, 1);
    BMI = zeros(num_students, 1);
    %Now we assign heights and BMIs for each student (based on RESOURCES
    % below). Note the multiplier in the height computation to convert to
    % centimeters. Then compute corresponding weight (formula based on CDC
    % resource). Finally, Basal Metabolic Rate can be computed with
    % height, weight and age.
    for i = 1:num_students
        if is_male(i) == 1 %i.e., student is male
            if grade(i) == 9 % avg height is about 64.5 inches
                             % and avg BMI is 19
                height_in_cm(i) = (64.5 + 1.5*randn(1,1))*2.54;
                BMI(i) = 19 + randn(1,1);
            elseif grade(i) == 10 % avg height is about 66.5 inches
                             % and avg BMI is 20
                height_in_cm(i) = (66.5 + 1.5*randn(1,1))*2.54;
                BMI(i) = 20 + randn(1,1);
            elseif grade(i) == 11 % avg height is about 68.5 inches
                             % and avg BMI is 21
                height_in_cm(i) = (68.5 + 1.5*randn(1,1))*2.54;
                BMI(i) = 21 + randn(1,1);
            else %i.e., grade = 12, so:
                             % avg height is about 69 inches and
                             % avg BMI is 21.5
                height_in_cm(i) = (69 + 1.5*randn(1,1))*2.54;
                BMI(i) = 21.5 + randn(1,1);
            end
            %Compute weight based on height and BMI: (eqn from CDC)
            weight_in_kg(i) = (BMI(i) * height_in_cm(i)^2)/10000;
            %Harris Benedict equation for Basal Metabolic Rate (BMR):
            BMR = 88.362 + 13.397*weight_in_kg(i) + 4.799*height_in_cm(i) ...
                        - 5.677*age(i); %this is the revised Harris-Benedict eqn
        else %i.e., student is female
            if grade(i) == 9 % avg height is about 63.5 inches and
                             % avg BMI is 19.5
                height_in_cm(i) = (63.5 + 1.5*randn(1,1))*2.54;
                BMI(i) = 19.5 + 1.5*randn(1,1);
            elseif grade(i) == 10 % avg height is about 64 inches and
                             % avg BMI is 20
                height_in_cm(i) = (64 + 1.5*randn(1,1))*2.54;
                BMI(i) = 20 + 1.5*randn(1,1);
            elseif grade(i) == 11 % avg height is about 64 inches and
                             % avg BMI is 20.5
                height_in_cm(i) = (64 + 1.5*randn(1,1))*2.54;
                BMI(i) = 20.5 + 1.5*randn(1,1);
            else %i.e., grade = 12, so:
                             % avg height is about 64 inches and
                             % avg BMI is 21
                height_in_cm(i) = (64 + 1.5*randn(1,1))*2.54;
                BMI(i) = 21 + 2*randn(1,1);
            end
            %Compute weight based on height and BMI:
            weight_in_kg(i) = (BMI(i) * height_in_cm(i)^2)/10000; %from CDC
            %Harris Benedict equation for Basal Metabolic Rate (BMR):
            BMR = 447.593 + 9.247*weight_in_kg(i) + 3.098*height_in_cm(i) ...
                        - 4.330*age(i); %revised Harris-Benedict eqn
        end
    end

%Estimate of daily calories expended:
    calories_expended = 1.1*BMR.*ALF


%%%%%%%%%%%%%%%%%%%%%%%%%%%%%%%%%%%%%%%%%%%%%%%%%%%%%%%%%%%%%%%%%%%%%%%%%%%%
% RESOURCES:
%%%%%%%%%%%%%%%%%%%%%%%%%%%%%%%%%%%%%%%%%%%%%%%%%%%%%%%%%%%%%%%%%%%%%%%%%%%%
%
% Male and female height charts:
%     www.heightweightchart.org/babies-to-teenagers.php
%
% Male BMI chart:
%     https://www.cdc.gov/growthcharts/data/set1clinical/cj411023.pdf
%
% Female BMI chart:
%     https://www.cdc.gov/growthcharts/data/set1clinical/cj411024.pdf
%%%%%%%%%%%%%%%%%%%%%%%%%%%%%%%%%%%%%%%%%%%%%%%%%%%%%%%%%%%%%%%%%%%%%%%%%%%%
```

Notice that this code has a nice header that explains the purpose of the code. Throughout, there are comments that explain the purpose of the different sections of the code.

REFERENCES

[1] SIAM M3 CHALLENGE PROBLEM, *https://m3challenge.siam.org/archives/2014/problem*, February, 2014.

[2] SIAM M3 CHALLENGE PROBLEM, *https://m3challenge.siam.org/practice-problems/2017-challenge-problem-sea-shining-sea-looking-ahead-national-park-service*, February, 2017.

[3] J. BIANCHI, NASCAR Atlanta 2017: Kevin Harvick costs himself win by speeding on Pit Road, *https://www.sbnation.com/nascar/2017/3/5/14824218/2017-nascar-atlanta-kevin-harvick-pit-road-speeding-penalty*, 2017.

[4] K. M. BLISS AND K. R. FOWLER AND B. J. GALLUZZO, Math Modeling: Getting Started and Getting Solutions, *https://m3challenge.siam.org/resources/modeling-handbook*, 2014.

[5] B. MÜLLER, S. MERK, U. BÜRGI, and P. DIEM, Calculating the basal metabolic rate and severe and morbid obesity, Praxis 90 (2001), pp. 1955–1963.

[6] NATIONAL OCEANIC AND ATMOSPHERIC ADMINISTRATION, Atmospheric carbon dioxide at Mauna Loa observatory, *https://www.esrl.noaa.gov/gmd/ccgg/trends/data.html*.

[7] NATIONAL OCEANIC AND ATMOSPHERIC ADMINISTRATION, Mean Sea Level Trend, Station 8536110, Cape May, New Jersey, *https://tidesandcurrents.noaa.gov/sltrends/sltrends_station.shtml?stnid=8536110*.

[8] NATIONAL OCEANIC AND ATMOSPHERIC ADMINISTRATION, U.S. linear relative mean sea level (MSL) trends and 95% confidence intervals (CI) in mm/year and in ft/century, *https://tidesandcurrents.noaa.gov/sltrends/mslUSTrendsTable.htm*.

[9] UNITED STATES CENSUS BUREAU, Population Division, Census Bureau reports, *https://www.census.gov/en.html*.

[10] WIKIPEDIA: Anscombe's quartet, *https://en.wikipedia.org/wiki/Anscombe%27s_quartet*.

NOTES

NOTES

NOTES